The 15 Best Discussions on Reddit Ever

The 15 Best Discussions on Reddit Ever

EDITED BY MICHAEL KOH

BROOKLYN, NY

THOUGHT
CATALOG
Books

Copyright © 2018 by The Thought & Expression Co. All rights reserved.

Published by Thought Catalog Books, a publishing house owned by The Thought & Expression Co., Williamsburg, Brooklyn.

Second edition, 2017

10 9 8 7 6 5 4 3 2 1

CONTENTS

Introduction—Michael Koh … 1

1. 40 Freaking Creepy-Ass Two Sentence Stories … 3
2. 33 Heartwrenching Last Words Of People On Their Deathbeds … 11
3. 50 OMG Stories From People On The Internet That Will Definitely Make Your Heart Stop … 29
4. 50 Unintentional Quotes From Children That Will Send Shivers Down Your Spine … 81
5. 50 People On 'The Time I Met A Celeb And They Acted Like A Total Weirdo' … 103
6. 66 Soul Punching, Evil Things People Have Done, Said, And Experienced … 123
7. 44 Employers Reveal The Most Ludicrous Things They've Ever Read On A Resumé … 173
8. 39 People On 'What Is The Craziest Sex Act You've Ever Participated In' … 183
9. 50 People On 'The Secret My Company Doesn't Want The Public To Know' … 197
10. 21 Pieces Of Life Advice From People Over 60 … 215
11. 50 People On 'The Secret I Am Terrified To Tell' … 227
12. 50 People On 'My Most Embarrassing Sex Story' … 251
13. 21 People Confess The Big Secrets That Could Destroy Their Relationship … 277
14. 30 Relationship Red Flags That Most People Ignore … 285
15. 40 People Confess The Little Things They Find Incredibly Sexy … 295

Introduction

Michael Koh

Reddit is a giant bulletin board system on the web where registered users can submit links, pictures and text ranging from the cute to the controversial. Posts can be voted up or down (Upvoting and Downvoting), which determines its position on the front page (or the subreddit).

Founded in 2005 by two graduates of University of Virginia, Reddit has become one of the most visited websites to date. They received their first funding from Y Combinator (think of it as the Yale of startup incubators) and in just over a year, Reddit was acquired by Condé Nast Publications.

Reddit is known for its diver and open user community. Categories are self-policed, with moderators flagging and removing posts that violate the Terms of Agreement or Service or not relevant to the subreddit itself. Due to its anonymous nature, it has spawned an incredible amount of little factoids delivered straight from the lives of the users, who all readily reveal themselves (truthfully or deceitfully, one can never tell) to take the weight off their shoulders, confess their minds, reveal parts of themselves that they never could in real life, for closure, for help, to leave a little part of their life on the web that they can never forget.

This book highlights some of the funnier, more provocative, soul-punching, salacious commentary provided by the users of Reddit.

1

40 Freaking Creepy-Ass Two Sentence Stories

I'm really scared to go to bed tonight. Found on r/AskReddit.

---

1. JUSTANOTHERMUFFLEDVO

   I begin tucking him into bed and he tells me, "Daddy check for monsters under my bed." I look underneath for his amusement and see him, another him, under the bed, staring back at me quivering and whispering, "Daddy there's somebody on my bed."

2. GAGEGE

   The doctors told the amputee he might experience a phantom limb from time to time. Nobody prepared him for the moments though, when he felt cold fingers brush across his phantom hand.

3. GRABOID27

   I can't move, breathe, speak or hear and it's so dark all the time. If I knew it would be this lonely, I would have been cremated instead.

4. ANARCHISTWAFFLES

   Don't be scared of the monsters, just look for them. Look to your left, to your right, under your bed, behind your dresser, in your closet but never look up, she hates being seen.

5. THEREALHATMAN

   I woke up to hear knocking on glass. At first, I though it was the window until I heard it come from the mirror again.

6. KNOWSGOODERTHANYOU

   They celebrated the first successful cryogenic freezing. He had no way of letting them know he was still conscious.

7. PGAN91

   She wondered why she was casting two shadows. Afterall, there was only a single lightbulb.

8. HORSESEVERYWHERE

   It sat on my shelf, with thoughtless porcelain eyes and the prettiest pink doll dress I could find. Why did she have to be born still?

9. BENTREFLECTION

The grinning face stared at me from the darkness beyond my bedroom window. I live on the 14th floor.

10. GUZTALUZ

There was a picture in my phone of me sleeping. I live alone.

11. MARINO1310

I just saw my reflection blink.

12. HCTET

Working the night shift alone tonight. There is a face in the cellar staring at the security camera.

13. MIKEYSEVENTYFIVE

They delivered the mannequins in bubble wrap. From the main room I begin to hear popping.

14. TUSKEDLEMON

You wake up. She doesn't.

15. CALAMITOSITY

She asked why I was breathing so heavily. I wasn't.

16. MADAMIMADAMIMADAM

You get home, tired after a long day's work and ready for a relaxing night alone. You reach for the light switch, but another hand is already there.

17. SKUPPY

My daughter won't stop crying and screaming in the middle of the night. I visit her grave and ask her to stop, but it doesn't help.

18. FLUFFYPONYZA

Day 312. Internet still not working.

19. ANONYMOUS_ABC

You start to drift off into a comfortable sleep when you hear your name being whispered. You live alone.

20. STORYTELLERBOB

I kiss my wife and daughter goodnight before I go to sleep. When I wake up, I'm in a padded room and the nurses tell me it was just a dream.

21. WAYSAFE

I needed to quickly run a SQL command to update a single row in an Oracle DB table at work. To my horror, it came back with "—2,378,231 rows affected."

22. HESUSMENDEZ

You're laying in bed and with your feet dangling out of the covers. You feel a hand grab your feet.

23. TLFMOD

The funeral attendees never came out of the catacombs. Something locked the crypt door from the inside.

24. THE_D_STRING

My wife woke me up last night to tell me there was an intruder in our house. She was murdered by an intruder 2 years ago.

25. AMMORTH

"Mesa called Jar-Jar Binks. Mesa your humble servant."

26. VIGRIDARENA

I was having a pleasant dream when what sounded like hammering woke me. After that, I could barely hear the muffled sound of dirt covering the coffin over my own screams.

27. SCRY67

The last man on Earth sat alone in a room. There was a knock at the door.

28. COBALTCOLLAPSE

After working a hard day I came home to see my girlfriend cradling our child. I didn't know which was more frightening, seeing my dead girlfriend and stillborn child, or knowing that someone broke into my apartment to place them there.

29. COMPARATIVELYSANE

You hear your mom calling you into the kitchen. As you are heading down the stairs you hear a whisper from the closet saying "Don't go down there honey, I heard it too."

30. EYEHATE

I was stoned. And Taco Bell was closed.

31. GENETICALLY_WITLESS

I never go to sleep. But I keep waking up.

32. ICHOKEDCHERYLTUNT

Nurse's Note: Born 7 pounds 10 ounces, 18 inches long, 32 fully formed teeth. Silent, always smiling.

33. AERRON

She went upstairs to check on her sleeping toddler. The window was open and the bed was empty.

34. BLAQKMAGICK

The longer I wore it the more it grew on me. She had such pretty skin.

35. VAULTKID321

"I can't sleep" she whispered, crawling into bed with me. I woke up cold, clutching the dress she was buried in.

36. DKMINO

You hear the scream across the hallway, but your eyes won't open and you can't move.

37. SCABBYCAKES

Being the first to respond to a fatal car accident is always the most traumatic thing I see as a police officer. But today, when the crushed body of the little dead child boy strapped in his car seat opened his eyes and giggled at me when I tried to peel him out of the wreckage, I immediately knew that today would be my last day on the force.

38. OWLLETTE

I looked out my window. The stars had gone away.

39. HANGUKBRIAN

I always thought my cat had a staring problem, she always seemed fixated on my face. Until one day, when I realized that she was always looking just behind me.

40. YOSHKOW

The pairs of emaciated eyes outnumber the single round in my gun. With pleading tears falling on her doll's hair, I point the barrel at my last surviving daughter.

Scared ya.

2

33 Heartwrenching Last Words Of People On Their Deathbeds

WHAT ARE THESE FEELS. WHAT IS THIS POURING DOWN MY FACE? Here's 33 stories of people's last words that will make you tear up. Found on r/AskReddit.

---

1. AWOD76

    Paramedic here.....I was transporting/treating at drunk 20 something who nailed a big tree. He was messed up pretty bad and drunk as hell. I asked him where he had been partying, to which he told me. I asked how was the crowd? He muttered, "So many hot bitches." He went unconscious, vitals tanked, and he died.

2. SMITHSKNITS

On her deathbed, my Yugoslav grandmother was hopped up on morphine to ease her pain. My dad and I were the only people in the room. She told us that when she was in the concentration camp, she and her sister were gang-raped by a group of Nazis while they were digging out in the forest. She said that they pushed her and her sister down into the hole and raped them for an hour or two. She was just kind of staring into the middle distance as she said it, so I don't even know if she knew that we were there. She ended her story by saying that she wound up pregnant and had to abort in the camp. She closed her eyes and slumped.

I actually thought she had died right then and there, but she woke a few moments later and asked if my dad had turned in her absentee ballot (this was during the 2000 election). Classic Baba.

3. UBERWOLF0

I was with my mother when she died. She was an abusive parent who did all sorts of nasty things to everyone around her, especially we 3 kids. In the end I was the only one who came to her when she was dying so she wouldn't die alone.

She said to me, "I wrecked your life." I said "What? no. You made me the person I am today. Aren't you proud of who I am?" she nodded her head yes.

It wasn't the very last thing she said. The last thing she said was crying out for my brother who never showed up until the funeral. I've never told anyone that part.

4. COMMERCIALPILOT

My great-grandmother went to bed one evening and didn't wake up for a number of days. Finally, 4 or 5 days later, she awoke, lifted her head, looked at her husband of 70 years marriage and said softly "I've loved you for 70 years now and I would do it all over again." Then she looked at her daughter and said "Daughter" nodded her head, laid back down and died in her sleep shortly thereafter.

Edit: Just to clarify, she didn't say "daughter" in a boring, emotionless monotone voice. It is a bit hard to understand since you weren't there to hear the tone of love in her voice and eyes. It was all very peaceful.

5. MAYASEYE

When I first started as a 911 dispatcher I had a call come in and all that the person said was "Tell them I'm sorry" and hung up. I knew right away what we were going to find when we got there. It was the worst feeling i just felt so dirty that I was the last one to talk to this guy, and no matter how fast we sent help it didn't matter it was just too late. So I guess he was confessing, but it just made me feel icky.

6. JACOBTWO-TWO

I was a health care aid on a geriatric ward when a woman so old and frail she looked dead already motioned to me to come to her. I put my ear next to her mouth and she quietly said, "I just wanted to say 'goodbye' to someone." It broke my heart. She died a few days later.

7. FATESARCHITECT

My cousin had Cystic Fibrosis, and had gotten a double-lung transplant at the age of 24. I'd grown up knowing that she was most likely going to die young, but with her lung transplant we thought she'd get another decade or two at best. About two months later I got a call late at night, saying that she'd been admitted to the emergency room, then the thoracic ICU. Unfortunately, she had bilateral pneumonia and a fungus in her lungs. My mom and I drove halfway across the country to see her, and it was awful. Her organs were shutting down, and her parents and doctors were debating whether or not to get her a second lung transplant.

She had decided not to go to college, instead trying to do different things in life, because I think she knew her time as an adult was limited. She went to Disney World (we were huge Disney fans as kids, especially Little Mermaid) with Make a Wish Foundation. She wanted to see the world, do everything she could. So when I came to see her, I sat and rubbed her cheek—pretty much the only spot on her body that wasn't bruised or had tubes running in and out. I told her about my recent trip to Africa, about elephants in our camp, about living in Scotland, about my recent semester in college. I told her, "I'll do everything for you. It's okay."

She opened her eyes and smiled at me, and then

closed them again. It was the last time I saw her awake and alive. She died a few days later; she got the second transplant, and never woke up.

She loved butterflies, and since she died, I've had them land on me with strange regularity all over the world. She's going with me because I'm living for both of us, or so I'm going to keep telling myself.

8. LLANDRYN

I'm not a doctor, nurse, etc., but I know of a death-bed confession. My mother, who is 90 years old, was born out of wedlock and given up for adoption. Her biological mother kept the birth a secret until she was on her deathbed. When she was dying she told her son, my mother's half-brother, that she'd had a little girl. They thought she was delirious, but later learned it was true. When my mother learned of this she cried and said "She never forgot me."

9. LOL_NICKELBACK

When my Dad was dying of pancreatic cancer, he told me that he had a son when he was much younger that he had to put up for adoption. My Dad was a pretty aggressive alcoholic in his earlier days so it was likely for the best. But he said he had no idea where he was now and regretted not ever meeting him as an adult.

10. ORANGE_PENGUIN

This may not seem like much, but the last words my

mom said to me before she died were 'Baby, I'm scared'. She wasn't scared of anything – she was a paramedic for over 20 years and had practically seen it all. That was the most horrifying part about watching her die. In all the times she'd been forced to go to the hospital, I had NEVER heard her say that she was scared.

11. PHILIPKDAN

My grandfather had another family out in Colorado (I live in NY). He had been suffering with Alzheimer's for a few years, and it was getting bad. Since he was technically my step-grandfather, his family from out west wanted to spend the last years with him. It really saddened my family, and brought all of us down, especially my grandmother, who never married him, but fell in love with him after he helped her through the passing of her husband and raising my mother.

He was an incredible man, but none of that was left in him. He couldn't hold conversations or really remember much. He had only begun to forget names. It was hard to watch, but in a way, I was grateful that I would get to remember him for the man he was before the disease.

As a freshman in high school, I had to understand that I was saying goodbye to him forever, even though he'd still be out across the country suffering. It was horrible, but it was necessary. In the last few days that he was here, he was very distant. In a way, he was aware of what was going on, but all of the details were fuzzy. He hadn't said my name in a while, but he was still clever enough to avoid having to say it or address me by name.

The day came for him to leave, and as he got in the car to go to the airport, he turned around and hugged me. I couldn't say anything, which sucked, because I knew that he actually couldn't physically say anything. He just looked in my eyes with the last bit of soul he had left.

Then, as he pulled away from the hug, he said in his frail dying voice: "I'll never forget you, Daniel."

And I never saw him again.

12. SUTURESELF8

Before I was a doctor, I worked as a nurse's aide at a small community hospital. There was a frail, elderly woman who lost her sight to cataracts. I took care of her for weeks, but she was suffering from dementia and couldn't really remember me from one day to the next. She was cute and harmless, mostly she would lay in bed staring at the ceiling, occasionally muttering nonsense.

One day, she happened to sit up abruptly while I was in there and screamed. I ran over to her and asked her what was wrong. She said, "Trevor... I'm sorry. I'm so sorry. Please forgive me" over and over.

I had no idea what to do, but she was hysterical. I sat next to her, took her hand, and said, "I know. I forgive you." She laid back in bed and seemed content. I left in the afternoon and she died that night.

I felt guilty for a long time for lying to her. I hope

somewhere, at some time, in some universe, Trevor really did forgive her.

13. PRINCESSTT

I watched my brother begin dying when he was about nine and every doctor said he wouldn't be alive to see 15. When we knew how serious his disease was, I told him I'd be there when he went. I let him down due to some complications with getting my paperwork done for emergency leave but my kids were visiting. When he found out their flight was delayed a day he said that he didn't know if he could make it and everyone just thought he was really excited (he adored my children.) He made it for five hours once they got in and then passed away at 24. He was a total badass. He also told my dad, a couple days before he went, that there was so much he (dad) didn't know about me and that, no matter how long he lived, he'd never tell. That man loved me more than almost anyone else ever has or will. My best friend to the end. I don't care if no one sees this. It's nice to just be able to talk about him.

14. VIVLAV144

I worked for a patient monitoring company and saw many people in many hospitals. I many near or at death. 99% of the time the y just wanted to talk and were always interested in what I was doing in their room. Much much more so than any other type of patient. I ended up asking an elderly man why so many people near death were so talkative this way. His answer was obvious to me only after the fact. He said he's dying and realized what he would miss most of all in this world... The stories. Took me awhile, thought a lot about it and realized myself just how powerful of a statement that is. His story, my story, and humanity's story.

15. GILLYBILLY

My dad's last words (heart attack in the shower, he died while we were waiting 45 fucking minutes for an ambulance to travel two miles) were: "I always knew I'd be bollocks-naked when I died, My life was fucking brilliant."

16. GENL

When she was in the hospital with terminal cancer, my mom confessed to me she cheated on my dad when they were still together. It wasn't a particular shocker. They had separated years earlier. I said "who wouldn't have? He was miserable and emotionally inaccessible." I think she felt better.

She also said "Of course I have to die of ASS CANCER. Everyone's donating money to breast cancer and wearing ribbons. Nobody cares about ASS CANCER." She was funny right up to the end.

17. RICHARDASTERD

My great uncle flew for the Navy in WWII and then flew commercial jets for 30 years. While laying in a hospital bed his last words to my dad were "Looks like a good night to fly."

18. SYDTHEDRUNK

My moms last words to me were: "You have to learn the difference between Chinese and Japanese people, because they don't like it when you mix them up." I wish I was joking.

19. IBOWDS

I'm not an EMT, nurse, doctor or anything like that. Just a regular college student. For the past 5 years I have been making different holiday cards for two retirement homes in the area. I make about 130 homemade cards for Valentine's Day, Easter, Fourth of July, Thanksgiving, and Christmas. I enjoy making them since a lot of people there don't have families, their families just dump them there, or you do have the lucky ones with families that visit. Anyway, I got a phone call from one of the homes yesterday telling me one of their patients passed away in the night. Her last wish was to tell me thank you for remembering her. She passed away holding all the cards I had made for her. This is why I'll never stop making them.

20. JALBERTS

My cousin's grandfather was in hospice for 11 days before he stopped breathing. His family was all at his bedside, and the doctor announced the time of death:

10:31 AM. At 10:38 AM he gasped weakly and then began breathing again for fifteen more minutes. The doctors didn't know quite what to do with that one...TOD is 10:31 AND 10:53.

21. SPHARTACUS

I had to tell my dad that it was ok for him to go, that I would be ok. A few minutes later he was gone. This was the last time I lied to my dad.

22. IMINYOURTOASTER

I don't know if this is relevant but on the night my grandpa died, he knew it was happening. He accepted it. For the last half hour of his life, he sat quietly while my grandma read him the love letters he wrote to her when he was a soldier in WWII. Prior to this night, he hadn't spoken a single word for nearly a week. Right before he passed he looked into my grandmas eyes, kissed her hand, and said "I love you". He couldn't have gone any more peacefully.

23. NORTHFOLKNATIVE

My grandmother was a pretty traditional Minnesotan grandma. The sweetest woman in the world, would never hurt a fly, and would scold anyone for saying "heck, darn, etc."

Until she had back surgery... at which point she was caught trying to sneak out of her room, thought my dad was her brother and told him to "go steal a car so they could get the fuck out of here."

Her actual last-ish words were at dinner the night after my grandpa died. She just quietly announced "well, I think I'm done then." She passed away that night. Terrific lady.

24. CHAKAKAT

I was once taking care of a dying man in his home. Coincidentally his wife was in the hospital also dying, making it a very sad time for the family. The man was feverish at night and would wake up confused and babbling at times. They had a freezer of soup the wife had made when she was well and the man was eating the last of it in his last days, but I don't think he even realized.

One night he told me a story of when he was in the war (I'm sorry I don't know which.) A paratrooper from the other side had landed in his territory and his parachute was caught in a tree. Against orders, when no one was around, he cut the man down and let him free. He asked the paratrooper to fire his gun when he was safe on the other side and let him run away, but was saddened to have never heard a shot. The man broke down telling me this and told me how he thought of it often and hoped the man had made it. He had never told anyone else. The man passed away a couple of days later. I've held this story close since.

25. GROOVEASSASSIN

My dad's a doctor and has a great "last words" story.

Back in the day, he was working/training for his residency at a nursing home. This particular day, he was working the late shift. At the nursing home, there was an elderly, short-statured man named Mr. Williams.

One night, around midnight, Mr. Williams left his room and walked down the long hallway until my dad found him. He was dressed in khaki pants and a button-down shirt and tie, as if he were heading out for the night or something. My dad asked him what he was doing, and why he was dressed up so nicely. Mr. Williams responded, "I'm meeting up with my wife tonight." However his wife had passed away seven years before that. He seemed to be deranged and really out of it, obviously. My dad and another physician brought Mr. Williams back to his room, and helped him get back into bed.

Around three hours later, my dad sees a senior medical director signing some documents, talking to three other doctors around him. He was curious as to what the commotion was, so he went over to talk. He asks the senior director what the papers were, and he replied: "Mr. Williams's death certificate…"

26. EXCIO

A daily occurrence during passing period was to jump from one concrete table to the one parallel next to it. After a few weeks of jumping every day the school had the janitor move it back a couple of feet. This was seen as a challenge instead of making us not want to do so. So my friend was the first to try. The point of the game was to make it from the table top to the opposite table top. In this case he didn't make it, he slipped on the part of the bench you sit on and crushed his throat. His wind pipe was completely broken and both sides were separated and pushed to opposite sides of the throat. Ill never forget the words he tried to mouth to me. "I can't breathe."

I have nightmares of that now. Its been almost 9 years. And this is the first time I have actually been able to tell people what happened.

27. [DELETED]

ER resident here. Guy came in with internal bleeding from blunt force trauma and looked at us and said, "Be excellent to each other." I shit you not.

28. QUEBECMEME

My grandfather's last words in his bed after years of painful cancer. As he sipped a beer he'd just requested almost inaudibly, but he took one sip of a brew through a straw. (He couldn't eat or drink so…odd request but we brought in a miller lite…) Then it made sense.

"These seats are AMAZING guys", with a happy tear in his eye and a slight smile at me, my mom, dad and uncle.

He was a diehard Bostonian, and thought he was at a Sox game.

29. REDDITNAME

I'm not a huge fan of revealing stories like that just because of the expected privacy that patient and families have.

But I hope you find this story interesting. There was one man who was on his last leg. We had been expecting him to pass for several days. His family was always around him and super supportive. They would joke around with him and me when I went to check on him. So we were starting to wonder if or when he was gonna pass at all.

One day, talking to the nurses, I learned that he had come very close to passing on several occasions, but his daughters would throw up a huge ruckus every time, and that would seem to bring him back.

Our Palliative counselor finally spoke with the family one day and told them many loved ones don't pass on if the family won't let them.

A few hours later, I walked by the room and I noticed that he was about to die. His heart rate was dropping. In my eyes, something beautiful happened. One of the daughters was about to start screaming and the other kept her quiet. They all stood there in silence until he passed.

Then they let loose the water works of poseidon!

30. ELR1804

When my grandmother died and her heart rate monitor flatlined, my grandpa got up to give her a kiss on the forehead, at which point her heart started and beat for a couple more seconds, and then slowly stopped again.

31. MRLINDERMAN

My mom has a huge fear of dead and dying bodies, so she couldn't bring herself to be in the room the last hour or so, but wanted to say goodbye to my dad in the last few minutes.

My dad flatlined before we could go get my mom, but I kept telling him that she was coming and wanted to say goodbye. After what seemed like a couple minutes (could have been a 30 seconds but it felt longer), his heart rate flickered and spiked for about two minutes. My mom got to say goodbye, and give him a kiss, and about a minute after she left the room, he went. He held on long enough for her to say goodbye, and long enough for her to leave the room, which was one of the most amazing things I've ever seen.

32. SKITT123

I was sitting with my grandmother in her bedroom. The doctors said there was no reason to keep her in the hospital, but rather to let her go home and die in peace. She had lung cancer and they kept her on a morphine drip. She was out of it for about 4 days and then while I was sitting alone with her, she suddenly sat straight up in her bed, looked at a corner of the room, on the opposite side of where I was, eyes wide open, mouth agape, and she said, "My god, you're so beautiful." Then her eyes closed and she slumped back into the bed. She died about 5 minutes later. Say what you will about the afterlife, but I think an angel was there to take her home.

33. NURSEGEEK

I work in pediatrics and the saddest confession I heard came from my six year old terminal cancer patient. She said that she blamed her brother one time for a bad thing she did. Her brother got in trouble and she thought God punished her with cancer and she thought if she set things straight he'd spare her. She died before she could apologize to her brother.

3

50 OMG Stories From People On The Internet That Will Definitely Make Your Heart Stop

Here are 50 ridiculously creepy stories from strangers on the internet. Check out 10 Creepy Videos, 51 Terrifying Mysteries and 40 Creepy Two Sentence Stories for more things to be scared about.

---

1. THATSAPADDLIN

In the late 70's, my Uncle was studying medicine at the University of Chicago. After a morning class, he decided that he would hitchhike back home to Lincoln Park on the North side instead of pay for a taxi. A man drove up in a Plymouth Satellite and offered my Uncle a ride. The man looked normal and seemed friendly...lighthearted even, so my Uncle got in the car and they started driving towards Lake Shore Drive. However, once they got there, the man drove South on Lake Shore instead of North, towards Lincoln Park. My Uncle told the man he was going the wrong way and to turn around and head North. The man looked at my Uncle, put his hand on his knee and said, "No son, you are coming with me" and smiled darkly at him. My Uncle froze in panic, and when they hit traffic near the South Shore, he quickly unlocked the passenger door and ran away without looking back.

A year or two later on a cold December day, my Uncle was having coffee in a cafe with my future Aunt when he caught something on the TV that made his blood run cold. He saw the man that had picked him up from school that day the year before. He had been arrested for the suspected rape and killing of over 20 young men and boys. The man on the television was John Wayne Gacy. And he had removed the door handle off the passenger side door to prevent the men he picked up from escaping.

2. EFFLUX

My family and I used to go "camping" a lot when I was younger. Camping consisted of renting a cabin in the woods and spending a little time in the wilderness. So we consistently rented this cabin in Pennsylvania where we would spend long weekends, when everyone in the family had some time off.

My two brothers and I, each being in the 9-12 year old range, would always run off into the woods and bullshit about while my parents did whatever. The cabin was on a mountain. If you followed a dirt road a ways past the cabin, the forest would open and there was a large field on the top. The field was about the size of a football field.

Near the edge of the field, on the far side, was a graveyard. The graveyard was pretty small, about 20 graves, surrounded by a wrought iron fence. The fence was about 10-12 feet tall with the gothic-ish spikes on the top. The fence had a gate but it was locked with a thick, rusty chain and padlock. Being kids were able to spread the gates apart enough to squeeze through.

The small gravestones were very old and worn, I remember seeing one dated 1890 something. On top of one of the graves, just resting on it, was a smooth black stone. It looked like Onyx or something, a little smaller than a golf ball but not perfectly round. My older brother pocketed it, we dicked around a little then left. Back at the cabin, which had one bedroom (where my parents stayed) and large living room/kitchen (where we stayed), we were hanging out while my parents were sleeping in bed. It was probably about 11:30 or so at night when a loud BANG! BANG! BANG! happened at the front door (which is right in the living room.) Me and my brothers were all scared shitless, understandably too afraid to answer the door. BANG! BANG! BANG! again the door shook moments later. It sounded like someone was trying to knock it off the hinges.

My father emerged from the bedroom asking WTF was going on. BANG! the door clashed. He knew by the looks on our faces we had no idea. He grabbed a wood chopping axe we had and walked over to the door. He looked scared shitless himself. He swung the door open and there was nothing but the night. No one in sight. After hounding us for information, and us having no idea, we went back to bed. I think no one slept much, if at all that night. The next day we were back to dicking around in the woods and we again found ourselves in the old graveyard.

The smooth black stone, that my brother took, was on top of the same grave. We ran, we ran fast.

3. ECHO5JULIET

I was driving a shortcut from Twentynine Palms, CA to Albuquerque, NM. Twentynine Palms is located in the desolate high desert east of LA. The shortcut was all two lane road through total nothingness, except for passing through Amboy, CA. Amboy is a nearly abandoned town nearly as far below sea level as Death Valley, with a dormant volcano and lava field on one side and a salt flat on the other. It was also, at the time, a hotspot for satanic group activity.

So I was driving by myself in the afternoon. I stopped in Amboy and snapped a picture of the city sign, just to prove I was there to friends who dared me to take that route to I-40. I got back in my car and proceeded to drive up into the mountain range between Amboy and I-40.

Once I reach the top I am driving north through a canyon with high grass on both sides of the road. Up ahead I see some stuff in the middle of the road.

As I approach I slow down to see a red Pontiac Fiero stopped sideways across both lanes, a suitcase open with clothes scattered everywhere and two bodies laying face down in the road, a man and a woman.

I stop a hundred feet or so away and the hair on the back of my neck is standing up. Being a Marine, I reach under the seat and pull out a 9mm pistol and chamber a round. Something seemed very wrong, it looked too perfect as if it were staged. An ambush? Was I being paranoid? Something was just wrong. Getting out of the car seemed unthinkable, it was the horror movie move.

As I scanned the road I saw a line I could drive. Pass the guy in the road on his left, swerve to the right side of the woman, behind the Fiero and I'd be on the other side. I dropped it into first gear, punched it and drove the line I planned.

I passed the back of the Fierro without hitting it or either of the bodies in the road. I continued forward a couple hundred feet and slowed down so I could breathe and let my heart slow down. As I looked up into the rearview mirror I saw that the two bodies had gotten up to their knees and twenty or so people emerged from the tall grass on either side of the road by the car and bodies.

At that moment my right foot smashed the gas pedal to the floor and did not let up until I had to slowdown for the I-40 east onramp.

I will never know what would have happened to me had I gotten out of the car to check on the bodies or stopped my car closer to them. Somehow I do not think it would have been good. Sometimes real life can be scarier than a movie.

4. SMPX

I was in Taiwan one year when I was younger, and had travelled to a busy night market (these are popular gatherings that usually operate in the evening). Nearby I spotted a sign for a netcafe in a 5-6 story tall building. Thinking I'd fire off some quick emails, I walked in the dark, small entrance of the building. The building was older and hasn't been well maintained, but it's not out of the ordinary in Taiwan. The entrance just had a dark hallway that led to a small elevator.

I pressed the elevator call button and entered. The elevator was uncharacteristically new compared to the building, but I didn't think much of it. Like some Chinese buildings, there wasn't a fourth floor (it's considered bad luck since "four" sounds like "death"), so it just read 1-2-3-5-6, which was usual. I looked for the floor the netcafe was at— 6th floor, and pressed the button. It lurched into action quietly and began the ascend. When it stopped, I figured it was my floor so I instinctively began to step out. Right before stepping out, however, the sight outside the elevator stopped me. It was pitch dark, only lit by the light in the elevator, it looked like it hasn't been occupied for decades, with some random pieces of furniture covered with white cloth or similar. It was a small building, so each floor were single occupancy, so I could see pretty much the entire floor from the elevator. Thinking I must have gotten the wrong floor, I checked the light (that indicates which floor you're on). Strangely, there was nothing, none of the indicators were on, but the floor button to the netcafe was still lit so I know I haven't gotten there yet. All this happened within a couple of seconds.

That's when I noticed a figure moving in the distance of the floor— it was not very visible but I could make out what looks like a person dressed in some kind of gown, moving slowly towards the elevator. I was thoroughly creeped out, so I started pressing the close door button. As soon as I pressed it, the elevator light flickered off. I am this close to pissing my pants, and it's actually kind of freaking me out thinking back to it. The lights flickered back on under a second and the door closed, the elevator jolted back to life. A few moments later it opened again to the netcafe.

I am beyond relieved at this point. I walked out immediately and sat down at a computer. After gathering my wits a bit, I walked over to the cashier's desk and told them what I saw. The girl working there listened and her face turned a bit ashen, so I asked her if she heard of similar.

She told me that she's never experienced it, but some coworkers and occasional customers have brought it up – basically, the building has 6 floors, and the fourth floor had a history. Apparently the floor used to be a hair salon of sorts, until one of the employees killed herself there for some reason. She slit her wrists over the hair wash station and died. The store continued operations despite stories of weird appearances— when customers got their hair rinsed the water would look a little red, like the customer was bleeding, little things like that, and a couple people reported seeing someone's figure walking away in the mirror. Naturally, the business closed down a few months later.

The building owner tried to re-rent the place out, but never had any luck. Most businesses are quite superstitious, and no one wanted to rent the fourth floor after someone had died in it, even at a very cheap price. Finally, after dropping the price to nearly nothing, a stationary supplies store wanted to rent. During the renovations of the floor, however, several accidents would happen. Tools would end up in strange places, a mirror from the previous business shattered when no one was near it, and finally a worker had his hand jammed between the elevator doors when it closed on him unexpectedly. The workers refused to continue working and finally, the business left and the building owner finally gave up and shut down the floor. He then had the elevator company come in to replace the panel so that the elevator could not go to the fourth floor.

Let me repeat that- the elevator was programmed to never go to the fourth floor. It doesn't even have a button. But for some reason, sometimes when people take the elevator, it would go to the fourth floor and the doors would open, and some, like myself, would see a figure walking around in the dark.

5. TUNGSTEN_MAN

When I was young my grandma came over to our

house to babysit me one night. Nothing unusual happened the whole night. But when my mom got home she checked the answering machine, and there was a message a few minutes long. The message was just my grandmother's and my voice laughing hysterically for the whole message.

Like I said the night was normal, and there wasnt a moment when the phone rang, or laughed hysterically for minutes on end.

6. CODEMECHANIC

My friend's dad used to haul shopping carts from Western Washington to Eastern Washington for repair, then bring repaired ones back. He did this driving a semi with trailer.

One summer my friend rode with him and told me about what happened on a stretch of rural highway.

They're on the road and my buddy is starting to doze off. Ahead in the middle of the road is a box. He dad says "Hey, want me to hit that box?" and he just kinda grunts and shrugs, then closes his eyes.

A few seconds later he wakes up because the rig is screaming to a stop and his dad is yelling something while he jumps out. He (my friend) doesn't know what the hell is going on. He gets out his side of the cab and looks back down the road where his dad is running. His dad is chasing and yelling at two little kids. The box is kicked up sideways. The kids were in the damn box. He swerved around it at the last second because he "felt weird" about the box. It was 14 years ago he told me about this. I have never driven over a bag or a box since then.

7. DUSKTILDAWN

My family used to rent a house in town along with my aunt and uncle when I was very young that we eventually moved out of because of very strange things that happened while we lived there, but the most memorable and final straw was the night my aunt was using the toilet and just happened to look down at this small hole in the floor that had been there since we moved in and saw a man standing in the basement looking right back up at her smiling. My aunt ran out of the bathroom and screamed for my uncle. After explaining to him that there was a man in the basement my uncle went and got my Dad and they both went down the stairs (the only entrance into the basement) where they found nothing but footprints in the dirty floor where someone had been standing and moving around under the hole.

8. AVALONHILL

My 4 year old daughter was supposedly asleep when I heard noised coming from her upstairs bedroom. I tried to listen but could not make out what was being said. I approached the room, and she stopped talking. Thinking I alarmed her I went into the room. At the time she was sharing it with her 3 year old sister. I walked in and saw the 4 year old sitting up in bed. I smiled and said is everything o.k.? She said fine, but her sister said they were keeping her up. I asked who? My 4 year old said sorry but that she was talking. When I asked her who she was talking to, my 3 year old sat up and said "the girl in the window, she said you were coming." After I shit a brick, I asked who the girl was and they both said a girl comes and stands in front of the window at night and talks to them. Not knowing what to say, I said o.k. tucked them in and hung around outside their door. The next day I asked about the girl. they said she came back but was mad! I waited a few days and asked again. My 4 year old said the girl in the window was still mad. I forgot about it for about a week, when my wife said, who are the girls talking to upstairs. Freaked out I ran upstairs and both girls were sitting under the window looking up. They turned and looked at me and asked if I wanted to meet the girl. When they turned around, disappointed, they said the girl left. It has been about 5 years since and I have not heard about the girl in the window since then.

9. [DELETED]

When I was growing up my little brother, who was three at the time, used to sleepwalk through our house at night. He'd walk down to the basement where I slept and crack open my door between 11pm-2am. He'd then slowly push it open and shuffle inside. When I'd ask what he was doing he'd always stare at the floor and say "Where's mom?" I'd tell him that she was upstairs. He would repeat "Where's mom?" Each night I would take him back upstairs and lead him back to bed where he'd fall asleep.

One night at about 1am I awoke to hear crying at the bottom of the stairs. I walked out to investigate and he was sitting on the bottom step. I asked him what was wrong and again he said, "Where's mom?" I told him she was upstair and we should go get her. "No," he said staring at the floor, "there's a bloody head following me"

"What??" I asked. He looked up from the floor, stared me right in the eyes, opened his mouth and let out the shrillest blood curdling scream I have ever heard in my life. It scared the living shit out of me. It was so loud that the whole family got out of their beds to see what was going on. After that I'd never ask him what he was doing downstairs, I'd just take him immediately back to his room.

10. DIGITALEVIL

I remember one time I was upstairs late at night getting a drink from the kitchen when I heard my little brother talking down the hall. I stepped out of the kitchen to see what he was doing and found him standing at the sliding glass door right outside his room, with the door open, talking to what I assumed was himself.

As I approached I heard him say, "NO! You can't come in here." Since I couldn't see well in the dark, I asked who he was talking to. He turned to me with his eyes half open and said he was talking to, "the man outside."

Freaked me the fuck out. I closed the door and put my little brother to bed then went and hid under my covers for the rest of the night.

11. [DELETED]

My girlfriend was living with her mother at the time and there was always this little kid from across the street who would just stand and stare at the house. One day her mom is going to get in the car and go to work when the kid asks who the old guy is that lives with them and why he never leaves the house. Her mom is pretty puzzled and asks what guy? (she is divorced there were no males living at the house). The kid looks up at the 2nd story living room window that he is always staring at and points and says the guy who is always standing there staring out the window. Kind of scared her mom replies that there is no guy that lives with them, she said the kid turned whiter than white and just turned and ran. after that it was like he did everything he could to avoid the house or even looking at it.

12. EMMBER

We'd been driving from coast to coast, and in Nebraska, we decided we needed to sleep. We pulled on to a side road and bedded down in the back of our pickup (It had a cap on the back and a mattress in the bed. Cool truck) Then we heard these bloodcurdling screams. It was a woman. This was before cell phones. We hear a man grunt and the screams stop rather abruptly. The next morning we heard about a murder in the area. I think we heard a murder.

13. REDDITNAME

My oldest daughter used to do this too except when she was in her swing. She would look at a blank spot in the corner and just start giggling and laughing and sometimes would have a little "conversation" with whatever it was she saw.

She had some strange fears when she was younger(she's 7 now). She was PETRIFIED of any toys that moved on their own. Remote control cars, little dancing chickens, a caterpillar thing someone got her that wiggled across the floor, stuff like that. And I mean petrified like she would claw and scream trying to get away even if I was holding her. She would have episodes where she said her head hurt and it was hard to breathe(took her to the doc, nothing was wrong) She was also deathly afraid of fire, even if it was on tv or in a picture.

She has come and asked me questions that no 3 or 4 year old should be asking, like "Mama you don't want me to die and leave you alone again do you?" and "What happened to my brown eyes?" Her eyes are blue.

My brown eyed mother died before she was born from lung and brain cancer.

14. MITCHELWB

I've told this story before on Reddit, but it's still creepy to me and probably fits here just as well.

I had a girlfriend in high school whose step father had sexually abused her when she was younger. She had told me about it so I knew. She also would talk in her sleep. And not just say random things, she would start dreaming, then start talking and you could talk back to her. She would talk to you but not wake up.

One afternoon, she fell asleep and started talking. She kinda started to whine, then said 'help'. I asked what she needed help with. She was telling me that she was in her kitchen and that he was in his room. He wanted to play the game. I told her that I was just outside, all she had to do was come outside and I would save her. She continued to whine, and told me she'd get in trouble.

I tried to convince her a couple times to come outside to me, but she wouldn't do it. She was really sounding stressed so I woke her up. It was really weird and I felt so terrible for her. I told her what had happened and she just cried.

15. KEEPINITHAMSTA

When I was a child my family moved to a big old two-floor house, with big empty rooms and creaking floorboards. Both my parents worked so I was often alone when I came home from school. One early evening when I came home the house was still dark. I called out, "Mum?" and heard her sing song voice say "Yeeeeees?" from upstairs. I called her again as I climbed the stairs to see which room she was in, and again got the same "Yeeeeees?" reply. We were decorating at the time, and I didn't know my way around the maze of rooms but she was in one of the far ones, right down the hall. I felt uneasy, but I figured that was only natural so I rushed forward to see my mum, knowing that her presence would calm my fears, as a mother's presence always does. Just as I reached for the handle of the door to let myself in to the room I heard the front door downstairs open and my mother call "Sweetie, are you home?" in a cheery voice. I jumped back, startled and ran down the stairs to her, but as I glanced back from the top of the stairs, the door to the room slowly opened a crack. For a brief moment, I saw something strange in there, and I don't know what it was, but it was staring at me.

16. ARGLE

This is the most unnerving story I ever heard, supposedly it is true. My best friend said it happened to his Uncle.

His uncle had a sugar shack, where he boiled maple sap to make maple syrup, and above the open vat of boiling liquid was rafters. The dude's young daughter climbed up into the rafters and was playing up there, and then fell right into the boiling sugar solution. The vat was huge, so she went right under, and she came up screaming, flesh blistering off her, completely horrific. The guy took the big oar like implement he used to stir the boiling syrup and used it to push her back down under the water, and killed his own daughter, because he knew to try to rescue her, she'd pretty much be 100 percent burned, and would die an even more gruesome death.

17. [DELETED]

I was super excited to get my first apartment. It was in an old antebellum house that was split into four units. Very cool place to live. However, every time I was taking a shower, I would get this overwhelmingly creepy feeling. Like somebody was watching me. Then the dreams started. I kept dreaming about this old lady in a pink nightgown. Sometimes she just looked frail and sweet, and she'd say that I should go with her. She never said where we'd go. Other times, the dreams were terrifying. Her eye sockets were empty. Her hair was greasy, stringy, and falling out. Her mouth was twisted in a tormented scream. And she'd frantically claw the air trying to grab me. The longer I lived there, the more menacing the dreams got. Also, the feeling of unease and the feeling of being watched in the shower increased dramatically. By the time we moved out, I couldn't close my eyes in the shower. It sounds silly, but I had this overwhelming feeling that I was going to die or lose my soul or something if I had my eyes closed too long. After moving out, I discussed all these weird feelings with a friend of mine who had recently moved into a house across the street from the old apartment. I was trying to laugh it off. He said that another friend of his used to live in the apartment above mine several years ago. An old lady died in what used to be my apartment. Nobody else wanted to live in that unit for more than a couple of months at a time.

The building recently burned down. The fire started in my old apartment. They still don't know what started the fire. Still creeps me out.

18. JANDALOFDOOM

So, this is how my grandmother tells the story.

It was 1933 and she was thirteen, living in the middle of Manchester, England. One night she got out of bed to go to the bathroom, and as she wandered through past the staircase, she saw her aunty standing at the top looking out the window.

Curious she trotted upstairs and stood next to see what she was looking at, but only saw the back garden and the alleyway out the back. She turned to ask her aunty what she was looking at only to see a nebulous, faceless figure staring back down at her. The figure then reached out her hands and gripped my young grandmother's face. The next thing my grandmother remembers is her older brother (about 27) running down the hall towards them, picking her up and carrying her into the nearest room.

She then spent the next week in and out of consciousness, eventually recovering, but now without a sense of smell.

Her family insist it was all a hallucination caused by a severe case of influenza, which is probably true, but my grandmother said she never felt safe in that house ever again. She moved to New Zealand about 10 years later and only ever returned to England, and that house, once before she died.

19. TUNAMUFFIN

I'm sure I'm a little bit too late for anyone to see this, but this is my family's scary story.

So, my grandparents lived on a farm in the middle of nebraska. They had just gotten married, moved in together, and had their first baby. The baby was only a few months old and needed to be watched, but it was early morning and the cows had to be milked. My grandfather couldn't have done the work alone, he needed my grandmother to help. The labor was easy and only took a short while to be finished and the baby, my aunt, had been fed awhile ago and was soundly sleeping. So my grandfather and grandmother both went to the barn to milk the cows, leaving my aunt asleep.

They finish milking the cows and my grandmother heads back to the house while my grandfather stays in the barn to do work. But when she approaches the house, my grandmother notices the door is ajar and swinging gently in the wind. She figures it is probably nothing, but is nervous just the same. She calls for my grandfather, who reluctantly comes to soothe her nerves. They enter the house together and hear the sound of the toilet flush just ending. Strange, yes, but farm houses in this area at this time had rather shotty plumbing, so while they become more nervous, they remain calm. They then pick up their paces and head towards the cradle where my aunt has begun screaming. The light, hanging down from the ceiling, is swinging violently as if it was just thrown on. My grandmother goes to pick up my aunt, noticing a black hair on her white gown. Both of my grandparents had white blonde hair and there is no one around who could own this hair. My grandmother becomes hysterical when my grandfather notices the latch to the attic is swinging, as though someone has just crawled up inside of it. He goes toward it, readying himself to open it. My grandmother lunges at him and convinces him in between her sobs to leave instantly, jump in the truck, and drive to town. Reluctantly, he agrees.

They never found out if anyone was in the house or not. However, a week later, Charlie Starkweather (not sure on the spelling) was found less than 30 miles away from their home. He was, i believe, the largest serial killer in America for a short time when these events transpired.

20. RGYAGRAMSHAD

I know this one pales in comparison to some of the others, but it's the only one I can be sure it's true. (Also, excuse my writing, it's late)

Several months ago, my cat went missing in the woods, and I had to look for him. It was late at night, and the moon was a thin crescent, so the only source of light was my flashlight.

I had seen my cat several times, but he seemed to be scared of something; every time I got close, he'd run further away. At a certain point, he got scared of something, and ran back towards the house.

I started to make my way back, and saw a man. He was just standing there, absolutely still. He had nondescript brown hair, and a tweed jacket on. I could make out all of his facial features except his eyes, where there just seemed to be shadow.

I called out to him, but he didn't respond. I then said "I can see you, you know" and was greeted with silence. I turned, and walked a few steps, and turned around. He was a few feet closer. I turned, walked some more, and looked again. This time, he was partially hidden behind a tree.

I didn't need any more warning, I booked it back to my house, where my cat desperately wanted to come in. I locked the doors, and sat on my couch until I calmed down.

Ever since that night, every few weeks, I hear a noise at my window. A slow, loud, rhythmic tap. There's no trees out there, and most of the nights it's happened, there's no wind. Every time, I've been to scared to look. Probably for the better, the last thing I want to see is a man with a tweed jacket, and no eyes.

21. DAMIDAM

There was a hunter in the woods, who, after a long day of hunting, was in the middle of an immense forest. It was getting dark, and having lost his bearings, he decided to head in one direction until he was clear of the increasingly oppressive foliage. After what seemed like hours, he came across a cabin in a small clearing. Realizing how dark it had grown, he decided to see if he could stay there for the night. He approached, and found the door ajar. Nobody was inside. The hunter flopped down on the single bed, deciding to explain himself to the owner in the morning. As he looked around, he was surprised to see the walls adorned by many portraits, all painted in incredible detail. Without exception, they appeared to be staring down at him, their features twisted into looks of hatred. Staring back, he grew increasingly uncomfortable. Making a concerted effort to ignore the many hateful faces, he turned to face the wall, and exhausted, he fell into a restless sleep.

Face down in an unfamiliar bed, he turned blinking in unexpected sunlight. Looking up, he discovered that the cabin had no portraits, only windows.

22. [DELETED]

This is a true story, told to me by a man who had been working as a murder investigator for over 30 years at the time. I was told this story after asking what the creepiest case he'd ever been involved in was.

This happened in northern Scandinavia in the late '80s, in a part of the country that is mostly covered in dense pine forest. On the highway between cities in this part of the country, you do come across the occasional villages and secluded houses, but there are stretches that seem to go on forever with only pine trees as far as you can see. A young girl, in her early twenties, was taking a motor coach home after being on a trip down south, presumably visiting friends or relatives. This happened just as winter was approaching, and it was freezing outside after nightfall. This girl lived in one of these really small communites that you pass along the highway, but during the bus trip she fell asleep and missed her stop. Looking at her watch, she realized that they'd passed it only recently, and that if she were to get off she would be able to walk back in approximately three hours. Either that, or get off in the next city where she didn't know anyone or had any place to stay. She explained all this to the bus driver, who pulled off at the next parking space and let her off. That was the last time anyone saw her.

Almost fifteen years later, long after the search for her has been given up, she is stumbled upon by a hiker. Her dead body was found tied to a tree, well over an hours walk from the road into the dense, almost impassable forest. The autopsy showed no signs of physical violence of any kind. Someone had just left her there, alive.

23. BABBITT86

years ago my grandfather was dying of complications from Alzheimer's. My little sister gave him a white stuffed bear with a pink heart on the stomach while he was in his death bead. When you squeezed the bear it said "I love you" in a pre recorded voice. He would constantly squeeze the bear and the voice made him smile. My grandpop had the bear in his bed until he passed away. Several days before he died my mother made him promise that he would somehow let them know he had "made it there safe." After he died we placed the bear on the mantel above the fire place. The family gathered shortly after his death to remember him. Just as we all sat down in the living room.

The bear started speaking on its own. "I LOVE YOU, I LOVE YOU, I LOVE YOU…"

24. 5THAPE

This story (story within a story) was told to me by a friend Mark:

During high school Mark was over at his friend's house, we'll just call him Steve b/c I don't know his name. They were hanging out in Steve's room after just hanging up a poster of Limp Bizkit's "Significant Other" album cover that Steve had just bought. Minutes later Steve's dad comes by and orders him to take the poster down immediately. When he asked his dad why, he says never mind why and to just listen to him and do it. Steve is pissed but listens to his dad and takes it down.

Later Steve asked his dad why he told him to take down the poster. This is the story he told him:

When I was young, your aunt, uncles and I had to all sleep on the floor in one room because we were very poor. One night someone walked into the room, bent over my head and asked me, "Give me your soul." I was so afraid I'd pretend to be asleep. This thing would ask me to give him my soul a few times then get up and walk out of the room. This happened every other night until I reached my early teens. One night after he was walking away I opened my eyes to see his face. What I saw was a thing with the body of a human, he wore a hood and underneath the hood was a face of a lizard.

A few years before you were born we threw a Halloween party at our house with all of our family. We ran out of candy so I drove to the local market to pick some more up. On the way there I was listening to the radio and they said to call in with your scariest story. I pulled over to the side of the road, called them on a pay phone and told the story of this lizard man over the air.

When I got back home I could hear someone crying. I went upstairs into the bedroom and saw my sister crying on the bed. I asked her what was wrong. She said, "I heard you on the radio. That happened to me too."

25. SMOWE

This story is not as cool as some of the others, but I believe this actually happened. It takes place in the 70's.

My father's friend is a forest ranger who also happens to be a competitive pistol shooter. Like, who can hit the most targets the fastest. That sort of thing.

He is driving with his wife, playing car poker with a local car club. Basically, the group comes up with objectives, generally places to drive to, and you are awarded a card for accomplishing them. The goal, of course, is to get the best hand. All of these objectives are completed over a weekend, and are done individually. That is, the drivers go their own way. This is not a caravan.

So, he and his wife are driving their convertible, I think it was a Triumph, along the back roads of Western Washington, near Concrete, WA. They come around a bend in the road and a large pickup dragging a good-sized log comes tearing out, perpendicular to the road, so that the log is now blocking their path. Putting the car in reverse and looking behind him, the husband sees another truck do the same thing, effectively boxing him, and his wife, in.

Each pickup bed is occupied by three to four rednecks. One of whom has a shotgun. One of the rednecks, clearly the leader, jumps down from the bed and walks over to the car. It being a nice day, the windows are already down. The redneck leader leans into the car and leers at the man's wife. "That is a fine-looking woman you got there." The husband's pulse races.

Being a competitive shooter, and a forest ranger, technically a peace officer, the husband carries a pistol under the seat of his car. While the leader of the redneck bandits walked up to the car, the husband slowly reached under his seat and pulled up his revolver, and concealed it under his legs. As soon as he heard the leader imply that these men were going to rape his wife, he grabbed the leader by the collar and put the pistol to his forehead.

"You will have your guys move that log in front of us or I will blow your fucking head off."

The leader was silent for a moment. The calculations running in his head, he said, "You can't get all of us."

"Sure. But you will be dead."

Self-preservation ruling the day for the leader, he motioned for the front log to be moved. The husband drove forward slowly, keeping hold of the leader's collar until he was clear of the obstacle. He made it past and got the fuck out of there.

26. [DELETED]

My grandmother told me this and I believe it just because of the way she tells this story:

She was 11 years old and on vacation at her aunts house in Turkey, Yozgat (small village called Sarikaya, which means "Yellow field"). It was early in the morning. Before she woke up, she dreamed of her aunt and how she was standing in the kitchen and staring out of the window with a cup of coffee, like nearly every morning. In the dream, my grandmother says, she said "Good morning." Her aunt would turn to her, smile and let her sit at the table, where they usually have breakfast and say calmly: "Tell your mum to wake me up. Don't wake me up yourself, ok? And if you see a brown bird, follow the bird." She awoke from this odd dream and told her mum to wake up her sister.

My grandmother had to leave her aunt's house the same day. After a long day at school (she was visiting a private school at this time, the kind where they give children uniforms: after the Ataturkian revolution this was mandatory) she was on the way home, where suddenly a chirping bird was in front her. She tried to pass the bird but it would just not leave my grandmothers path and soon would chirp louder. At this moment she finally noticed that the bird had a brown color and she remembered what her aunt was saying in the dream: "Follow the bird." The weird thing was, she said, the bird wasn't flying at all. It just was jumping away from her and she soon followed it, till she reached the door of a police station, where an officer just opened the door and see the little girl which was my grandmother. The officer but soon would look behind her, to see a person with a black pouch accompanied with some kind of farmer truck waiting for this odd person. Needless to say, those persons have been kidnappers, which decided to flee immediately. The officer decided to bring her home that day. Eventually, she found out that her aunt died. Most likely at the day she left her house.

English is not my first language, but I wanted to share this with you.

27. DECORATOR

I married my college sweetheart right after we graduated. After about a year, it wasn't going well, and it seemed like it would be best for me to move out.

My brother, who is a few years older than me, lived on the other side of town, where he has a fairly large house. He is actually planning to move to another city, and staying there in rented accommodation pretty much all the time. He agreed that I can stay in house until I get myself sorted out or until he sells the house, whichever comes first. His house was actually kind of run down. He had been planning to fix it up, but was so busy with his job that he never really got that much done. I had plenty of time on my hands at evenings and weekends, so I volunteered to do some fixing and decorating for him.

Over a few months, I repainted all the rooms, fixed

all the wooden floors, and even retiled both the bathrooms, put in a new toilet and shower, etc. My brother paid for all the materials (he gave me a prepaid debit card), but I did all the labor for free of course. The last room that I was doing was a bedroom. It had a built-in wardrobe cupboard, kind of built into the wall. I decided to paint the inside of the cupboard as well as the room itself, since the cupboard is dirty yellow inside with lots of black marks on the walls. I used the last of the white paint to paint the inside, and left the doors open for it to dry. That was my Friday night, then I went to bed.

Next day was a Saturday, and the last thing to do is paint the walls of the room, which include a dark red lower half, and a cream upper half (there's a rail between them, and it didn't look as horrible as it sounds). I went to get the red paint from the corridor, where I had been using it too touch up a spot that I had missed. I then went back to the corridor to get the red paint tray with the roller and brush in it. I tripped as I entered the room, the tray and roller fell on the floor (which fortunately was covered), but the brush went into the cupboard and hit the wall. It left a mark that looked like an elongated S with a long line going straight down underneath it. Now I was pissed, because I would have to repaint the inside of the cupboard – at least a couple of coats to cover the dark red – which means I will have to go out and buy more white paint as well.

I picked up the brush, and start to write SHIT using the elongated S for the initial letter. The H however came out looking more like an A, so I write SATAN instead. There was still a long line of paint running vertically under the S, so I made that into the vertical stroke of a K, and wrote KILL.

I thought nothing of it, and then got on with painting the rest of the room. I spent several hours painting the entire room, and by the time I was finished, it was dark and late, and I was aching and really hungry. I decided to go downstairs to get some food and then go to sleep. As I was leaving the room, SATAN KILL caught my eye, and for some reason I decided to write in ORDERS YOU TO after SATAN, making the message SATAN ORDERS YOU TO KILL. It didn't seem important, as I am planning to paint over it anyway.

First thing Sunday morning, I went out and bought a tub of white paint. When I got back I paint over SATAN ORDERS YOU TO KILL, but you can still read it through the white paint. I then started on the second coat on the room proper. When I finished them room, I redid white roller over SATAN ORDERS YOU TO KILL in the cupboard again, but you can still read it. For the next week, every morning before I leave for work, and when I get back from working in the evening, I rollered another layer of white paint over SATAN ORDERS YOU TO KILL. I was convinced that it was still faintly visible.

The next weekend my brother came over, so I showed him the cupboard, and asked him if he can see any message written inside it. He said that he couldn't. But I was still convinced that is was faintly visible. I told myself that it is my mind playing tricks with me, and that I must take his word for it. Nevertheless, just to be sure, I did add a few more layers of paint over the next few days. During this time, there are periodically people who come with the realtor to look at the house. My brother was after all trying to sell it. I do particularly remember one family (mother, father and teenage boy) who spent ages looking over the house one Saturday – I think (not sure) if this is the same family that reappears later in this story.

I soon moved out, and moved away to another town. Got a new job, rented my own place. My brother eventually sold the house. I met a new girl, etc. At Christmas, my brother invited me and my girl over to his large apartment in a major city. We went to visit. When we are talking, he tells me that he is so glad that he is rid of that house, since it always gave him the creeps. Asks me if it ever gave me the creeps. (It didn't). Then the killer revelation: The family who bought the house – the teenage son killed his parents, and hid their bodies in a cupboard.

28. HIGHONFIRE

When I was growing up I lived in rural alabama, about 30 miles north of a place called Mount Pinson. There really wasn't anything around. Just houses, nothing like a neighborhood in the traditional sense. No stores nothing. We had a one police officer and he was pretty friendly. He was my best friends uncle. He would often let us ride around with him because nothing really ever happened. There wasn't a recorded murder or anything beyond someone being drunk and a public nuisance. He didn't even arrest people for that, he just ferried them home.

Well we had this lady who was all alone.. Her husband and died a few years earlier and she was lonely. She called this one cop over everything. If some kids were in her yard playing, she'd call the cops saying she was being harassed. She did it because she was lonely and wanted company. No one could figure out why she just wasn't nicer to the kids or anyone around the neighborhood. One day there was this new kid kicking around near this little spot we liked to hang out, it had a nice climbing tree and a rope swing. There wasn't any grass, just dirt. It was fun to kick the dirt up and stir it up and get all dirty.

My and my friend being interested in this kid introduced ourselves. He told us stories of living in a big city and how much fun it was. We were mesmerized at the time. He taught us some new games to play. We generally had a good time. Well his father was a real prick. It turned out this kid was the mean old ladies grandson and his father had lost his job and went back to say with his mom while he recouped and stuff. Since the kid was around she had warmed up a little to us and would invite us over for snacks and stuff. We obliged being kids we were always down for some free treats. One day the new kid never showed up to our little spot. So we figured he was busy with his grandma or was doing something else. A couple of days went by and my friend told his uncle. His uncle said he'd take a look. He let us tag along because he never imagined anything weird happening.

It took about 10 minutes to make it up the windy dirt road to her house. It was quicker just to walk through the field. Took maybe two minutes to do. Well, we get up there and he instructs us to stay by the car. We oblige him. He knocks and knocks. No answer. Their car is there. He knocked again this time louder. So he circled around to look in the back. My friend and I decided to try knocking. It didn't seem like a bad idea at the time. So went up to the door and knocked. Nothing. My friend grabbed the doorknob and turned and I was like, what are you doing?! And he pushed the door open. We both almost threw up instantly. It was like the smell was a physical force and pushed up against us. It was so vile.

Friends uncle made his way around and when he got to the porch and he took a smell he knew immediately what had happened. He instructed us to go to the car and wait. He went in and looked around and came bolting out of the house and threw up. He had tears in his eyes. Not like he was crying, but he just so upset. He went to his car to contact someone. My friend took off and I went after him. He was intrigued and so I was I. I felt a little more bolder following him in. We tip toed in… the first room was completely clean.. we went into the kitchen…

The kid was stuffed into the oven.. burnt.. he had clawed at the glass front of the door and actually scratched it. You could see his hands. His face. It looked like he was still alive trying to get out.

My friend and I just sat there. In complete shock for what seemed like eternity. It was so surreal. My friends uncle snapped us back to reality as he came running in the house and grabbed us, screaming yelling, telling us never to disobey him again like that. He took us both outside.. we just sat there… the entire evening we didn't say a word to each other or anyone.. My friends uncle dropped me off. I never saw my friend again. Uncle said he couldn't handle it and went to live with some family members in the city.. My family moved a year later. I've never been near that house again.

Uncle ended up drinking himself to death and was depressed.

A farm that was near the house ended up having some sort of infestation and they lost all their crops. This weird circle of death expanded around the house.

29. OBSCENEBIRDOFNIGHT

I have Google Voice set to transcribe any voicemails I get and send them to me via text message. So, I get a call from an unknown number, so I let it go to voicemail and the transcription that it sends me via text message is "Hi, Fritz."

So later that day I check my Google Voice page and listen to the voicemail. It is 30 seconds of absolute silence. Google Voice transcribed the block of silence into the greeting "Hi, Fritz".

The creepy part about it is, the only person who called me by my nickname of "Fritz" was my grandfather, who has been dead for 3 years.

30. ERIS_AMAZING

I grew up in a tiny, tiny town in Northern Ontario. Every Friday, most of the neighborhood would gather in Ginette and Emil's kitchen and tell stories for hours. I was only five or six, the only kid on the block, so I'd usually read or play with my toys in one of the two guest bedrooms until I fell asleep. One night, I couldn't sleep because I could hear a baby crying. I came out into the kitchen, and complained to my mom. She told me to go back to sleep, but Ginette, an experienced grandmother at the time, pulled a cookie out from the cookie jar and sent me to the living room to watch TV instead.

Because of the way the house was laid out, the living room was right next to the kitchen, and I could hear the adults talking. I'm 22 now, in Texas, and it STILL creeps me the fuck out.

"When we first moved here," Ginette started, "There was a cross hanging up in the guest room. We're not really religious, and there was a lot of other crap from the old owners hanging up so we took it all down."

Their bedroom was in the basement, closest to the heater, and the kitchen, living room and guest rooms were all upstairs. "We used to hear something running around upstairs, like a kid, eh? We thought it was just the dog, or the cat. But one weekend, Emil took the dogs hunting and the cat wouldn't come in that night. So I was downstairs, alone, and I heard the footsteps. I came up to investigate, and I could hear a baby crying."

Now, the neighbors were too far apart for it to be one of their kids. And their house backed up into a field that went on for miles and miles. She was confused, and wound up calling the cops because the baby just would NOT stop crying. The OPP came, a grizzled older guy and his partner, and the older man remembered the house from when he'd been called there three years ago- when the Harrison's toddler was found dead in its crib. They searched the yard and the house, but didn't hear anything. Offhandedly, they suggested she find the cross, and put it back on the wall.

She did. And she never heard the baby again. When my mother went in to get my stuff so we could leave, she realized that when I was playing, I'd knocked the small cross off the wall.

31. MIKEQUIRK1

Once, when I was a teenager, I was waiting at an abandoned gas station in downtown Akron to meet a dealer to buy some weed. This was in about 1993 or 1994, so payphones were still functional and in pretty common use. As I was waiting, the payphone in the parking lot started ringing. Bear in mind, it was at about sunset on the outskirts of downtown and not another single person was around. Out of curiosity, I picked it up. The man on the other line asked, "Is this Chad?" My name isn't Chad so I said no. The man ignored me and said, "Chad, I want you to do bad things to me." I stated again that I wasn't Chad and asked him what he wanted, if he knew where he was calling, etc. He ignored me again and went into very explicit and specific detail about all the things he wanted Chad to do to him sexually. I was laughing and told him again that I wasn't Chad. Finally, he said he knew for sure I was Chad and described to me what Chad looks like. He described me perfectly down to the color of my shirt and what type of shoes I was wearing. I immediately hung up and looked around. There was nobody, I mean not a single person, around. I got into my car and got the fuck out of there.

32. MISSCRYSTAL

I was 7, lying in my bed, reading, when a woman I had never seen walked up and sat on the edge of the bed. She wasn't scary at all, in fact, she somehow made me calmer. She said "Don't look out the window. Just keep looking at me." Then she started singing a song I had never heard. She kept singing it over and over, quietly, and petting my hair. Every so often she would stop and tell me not to look at the window.

A few years later, I saw a picture of the woman from that night at my grandmother's house. It was her mother, who died when she was 10. She had just gotten the photo from her step-mother, who had been going through her father's belongings. Before that, she had never had a photo of her mother. I'd dismiss it as something I saw at Great-Grandma Dixie's house instead and had a dream about, except that I never in my life went to Great-Grandma Dixie's house. Any time I saw her I saw her at a church or a family picnic. A few years after that, I heard the song again. The song was Molly Malone, which I had never heard in any other context. I still remembered the words. I asked my grandmother if she had ever heard the song, and she had. It was the song my great-grandmother used to sing to her children to calm her down before she died.

When I told my mother about it, she asked when it was. Since it was a few days before Easter when I was seven, I was able to give her a vague idea. Turns out, right around that same time, her girlfriend was having nightmares that a ghost was floating outside my bedroom window, trying to pull me out. Much like you, I've always had weird things happen to me, and this is just the most vividly remembered and bizarre.

33. MEEKLEBERRY

A long time ago my dad was deer hunting in the mountains. Practically in the middle of no where. He was staying in this campground and a family of four invited him to eat dinner with them. While they were eating this other hunter that the family invited to eat kept asking my dad if he wanted to hunt with him tomorrow. The guy seemed kind of skrewy and my dad didn't trust him so he kept declining the guys offer. After dinner the guy kept up "Are you sure you don't want to go out with me tomorrow its supposed to be a good day." My dad got extremely creeped out by this, threw all his gear in his truck and got the hell out of there. A month or so later in the news rangers found a bunch of hunters bodies in that area. Apparently someone was killing hunters in the middle of the wilderness.

34. LEONGREY

One night I was having this really odd dream. It was one of those ridiculously long dreams, that seems like it lasts for years while you are asleep. This one lasted a whole lifetime.

I watched a woman live her life. I watched as she was a child, as she grew up. I watched her go through school, college. I watched her social life evolve, I watched her romance and her marriage and her pregnancy. I watched her live her mid-life, raise her children and then grow old. It was a pretty normal life, I didn't see any trauma. I didn't remember how she died, but the moment it was over I woke up.

I had to go to the bathroom really bad. I stumbled out of bed, probably 4 in a morning in a most-asleep daze and went to the bathroom. As I sat and peed, I saw her face looking at me through the window. There she was, old, wrinkled, wide eyed and gaping mouth with the most horrible, scornful face looking at me.

It didn't phase me at the time. I got up and walked back to bed and fell right asleep. I didn't realize until the next morning the horrifying thing I had seen.

35. PLASMITE

When I was 8 my mom, dad, and younger brother were visiting some relatives. We ended up staying the night, with my brother and I sleeping on cots near my parents bed.

In the middle of the night, I distinctly remember feeling incredibly cold and waking up. While attempting to rearrange my sheets, I looked up and saw my dad wide awake and staring intently at something behind me.

When I turned to see, I saw several small little glowing balls floating down towards my brother and I. Turning back to my dad, we made eye contact and I jumped up into bed with him, but he did not move or speak at all. Even under the comforter I was chilled to the bone.

He was visibly tense and concentrating very hard, and I swear I could see the little glowing balls get pushed away from us, but it wasn't quite enough before they reached my brother.

When they did, I didn't feel as cold anymore, but I felt very strange, like a mix of dozens of emotions without being any particular one. I looked at my dad, who looked very sullen, who put his arm around me and went back to sleep.

The next morning, we were awoken by my brother throwing up all over the carpet. This continued for hours, he had a terrible fever, chills, and had a really hard time speaking. My father moved him downstairs and gave him some pain killer, but my brother said that it didn't work.

About lunch time, we decided to take an early leave and go home. Less than 20 minutes into the car ride, my brother was completely fine again.

My father and I talked about this a few years ago, and only more recently told my brother and mom about it. We have since slept there in different rooms without incident, but my mom sleeping in there once on a trip had a similar experience (vomiting, fever, chills) only to be fine minutes later after leaving the house.

We don't sleep in that room anymore.

36. THEGURL

My grandmother, in Italy, during WWII, went to market and bumped into a friend of hers. They talked briefly, her friend seemed distracted and in a hurry. Later that day, my grandmother was talking to a neighbour and told them that she'd seen her friend on the way to market earlier that day.

"That's impossible," said the neighbour. "She died two days ago, the funeral was this morning."

Yeah, my Nonna bumped into her friend on the way to her own funeral.

37. RON_MEXICO_MD

One night, this guy was driving back home from his girlfriend's house when a large dog ran out in front of his car. My friend had no time to swerve and hit the dog dead on. Being an animal lover, he jumps out of the car to see if the dog is alright or not. But before he can come around to the front, he sees the dog get up on two legs and run off into the field next to him.

38. INCONSPICUOUSLY_HERE

My brother went through a pretty nasty divorce after two years of marriage, and having two kids with the girl. My niece was at her day care, and she says, without being provoked, "my mommy is dead in the woods." Well of course the day care workers are very confused by this and call my brother to come talk to my niece. He asked her why she would say something like that, and my niece replied, "because i saw her there." The creepy part is, her mother ran away about a month before this happened with a very, lets say questionable guy, and we still have no idea where the fuck she is. This happened about a year ago, my niece was four at the time.

39. ENKEN90

I used to live in a house next to an abandoned train station, and one day a friend of mine and I decided to explore it. We had obviously done this before, but this time something was different. One of the old train wagons standing there suddenly had curtains. Naturally I tried to peek inside, but as I got close to the windows, the curtains suddenly flung open and an elderly woman stared back at me with a terrified look in her face.

I have honestly never been more scared in my entire life, and naturally I just sprinted the heck out of there. To this day I have no idea why that woman was inside there and I do not know what happened to her. Years later I walked past the train station at night, and I could see a TV with static standing inside the building that used to be the station office. Creepy place, never been a fan.

40. [DELETED]

I baked sourdough bread for years as a hobby. I'd make more than could eat and give the loaves away to friends and family.

When I moved to Albuquerque for work, I started baking bread again not long after. With no friends or family to fatten, I gave the extra loaves to boys living in the condo next door with their mother. I'd probably given them a dozen loaves over a year or so.

I came home from work one day to find the entire police department camped in the parking lot, including one of those large crime scene labs-on-wheels, yellow crime scene tape, etc. I asked what was going on and was told there had been a homicide and that I would have to sleep somewhere else for the night.

I went to a hotel, tuned in to the local news, and learned that the mother of three boys mentioned above had shot all three of her children in the head with a .38 caliber handgun, one by one as they came home from school…

…and she used loaves of bread to muffle the sound of the gunshots.

41. [DELETED]

My uncle died alone and was kind of an outsider/black sheep of the family. My dad and I went to his farmhouse in Oklahoma to inventory his stuff and begin clearing it out to sell the place. We go to the barn out back and open it up. About five paces in, a pitchfork flies out of nowhere and sticks into the wall next to my dad, barely missing him. I'm immediately looking for a place where the thing could have been resting, such that it fell when we opened the barn, or some sort of mechanism that could have flung it at us. I find nothing. Then I think maybe someone was in the barn, we startled them and they chucked it at us. Again I find no signs of anything.

Then I notice that my dad is eerily calm and quiet, seemingly not even interested in figuring out what just happened — but he's as white as a sheet. I think he's having a heart attack or something, and I ask him if he's ok. Bear in mind he's a scientist. He believes in nothing supernatural, but he proceeds to tell me that there is a family secret that I don't know involving this barn, my uncle, and a pitchfork. And that I must never bring the matter up again. Ever.

42. NATEZOMBY

I was in Bermuda in the off season (I think) with my Mom, Dad, little brother. I must have been about 5 or so, it was before my other brother was born and my sister was born.

I remember the hotel was empty and a man with a red vest/shirt got me some milk (I think).

I remember the beach was empty. Just so you know, most of the place was empty. Cheaper maybe for us to go?

I was on the beach with my parents and like 2-3 year old brother was with them. I ran off to explore. I saw my mom on the beach. "Come on Nate!"

There was a thicket area of scrubby bushes and trees along the beach. She went up into it. There was this cement square with pipes or vents or something coming out of it. She stood there, then saw me coming and kept walking into the thicket. I felt something was wrong, but didn't want to be defiant so I pretended I was really interested in the pipe things and said aloud "I wonder what these are for?" or something. So if she got mad I could say I was interested in the pipes or something, not just disobeying.

As soon as she went out of view into the thicket, my Dad came up from the beach into where I was and yelled "Where have you been!? Don't wander off like that!" or something similar.

He took my hand roughly and lead me back the beach where my mother and brother were.

I still don't know what the fuck happened. The woman looked exactly like my mom but felt wrong...

43. BLUELABEL

I was living with my girlfriend at the time and it was about 10 years ago. One night we were in bed and i woke up and i saw standing over my girlfriend a young man, i shit myself and jumped over my girlfriend and took a great big swing at this man standing over the bed hoping to knco him out by surprise. All i suceeded in doing was falling off the bed and waking my girlfriend when i hit the floor.

She asked what the hell was i doing, and i told her about the young man standing above her. She asked to describe him, to which i did and she replied, thats my friend who was killed a couple of years ago. He was a lumberjack and a tree fell on him while at work.

She showed me a photo the next day. Sure enough that was the guy standing over our bed.

44. ABSOLUTELYAMAZED

When I was 14 I was sleeping over at my friend Dave's house. His grandma was there watching us because his parents were out of the country. When it was time to go to bed his grandma gave us both a hug good night. When she hugged me she grabbed my ass and slipped her tongue into my ear. Creeped me out so bad I don't think I slept all night and never again at Dave's house. Fucking creepy, man…

45. ZANDERF87

One night, as we are watching tv, our dog starts to bark, going crazy at the door. Much more so than he ever does, which is MAYBE once every couple months. But he was really different about it this time. Fast forward two weeks, cops are combing through the small forest that is behind our backyard fence. Turns out a woman was raped and brutally murdered there.

46. BKH

A friend of mine related this story to me:
He and his wife decided to take their three young kids to Mexico on vacation. They're the type that want to see the 'real' Mexico, not the tourist towns. They talked to some people they knew from the area they were interested in (sorry, can't remember) and were

told that it's a great idea, just don't drive from city to city at night.

So they drove down and heeded the advice to only travel between cities during the day for a couple weeks.

One day, they realize that the city they want to see next is going to involve an overnight drive unless they take a flight and rent another car. They decide that nothing bad has happened so far and to just go for it.

It's well after sunset and the road they're on has gone into a canyon and the canyon is getting really twisy, the road really narrow.

They come around a sharp turn in the road and see a massive boulder in the center of the road. With some amazing luck he manages to avoid crashing into this thing and goes flying around the other side of the rock.

There's about 20 people just standing there looking at them, all with various farm implements and weapons at the ready. They also notice a car that obviously didn't avoid the boulder in the road, and bodies on the ground next to it.

47. KRYPTUS

In 2004, there was a girl I really liked. One night I summed up the courage to ask her on a date. We went to a restaurant, drank some wine, and at about 11pm we decided to go to a beach called Mokapu, which is near Honolulu. It was really dark and there was hardly anyone out. We talked about each others lives and experiences and we laid down and listened to the ocean for a few hours. At about 2am strange things started happening. Me and the girl I was with felt our hands being touched by something small. We thought it was miniature crabs, but when we looked, there was nothing there. The moon was full that night so you could see a lot of stuff, but we saw no crabs. Then the weirdest sound was heard by both of us. It sounded like a baby crying. Keep in mind it was 2am and no one was around. The nearest house is about 5 miles away. We kept on hearing the crying. We both thought that it was cat crying or something, but when we looked at the direction of the crying, there was no cat to be found. We kept on feeling our hands being touched too. We both had had enough and we decided to leave. I just put it to the back of my mind thinking that it was crabs and a cat.

Two months later I was sitting with my hair stylist, who is a local, and we began talking about strange things that happen on the island. We were talking about Night Marchers and haunted burial grounds. Then she started talking about Mokapu. I asked what is so scary about that place? She said that a long time ago, Hawaiians used to take the deformed babies and kill them at Mokapu. She told me some people say that if you stay there long enough you will hear them crying. She told me that she heard them herself one night when she was there. The hair on my neck and on my entire body stood up. I told her what happened to me there.

48. 911_THROWAWAY

On the night of September 10th, 2001 I had a couple of

very vivid dreams. The kind where you are half awake and almost daydreaming.

In the first one I don't remember how it started, but I saw an airplane crash into a huge metal structure with a lot of windows. To be honest I thought it was a train, but I was confused because it wasn't lying flat on the ground like a train. I then heard screams and they were so real to me that they woke me up. I was sweating and panicked. It took me about an hour to calm down and fall asleep.

Then, I had the exact same dream again and it woke me up again at the exact same moment and those screams sounded just as real. This time it took me almost 2 hours to fall asleep as I was deeply disturbed that I would have the exact same dream so vividly twice in one night.

When I finally fell asleep again I had one last dream. There were a bunch of people in the air and they were flying. Not in a plane, but just flying with their arms out like superman, zooming all over the place. Everyone was smiling and having fun. Then, all at once they stopped and started to fall. They all started screaming and I could see their terrified faces as they fell. Once again I woke up from the screams. I didn't sleep well the rest of the night.

The next morning I slept in (I was in college and only had late classes that day). At about 9:30 I was woken up by my girlfriend calling me to tell me about the plane crashing into the world trade center. I honestly had never seen the WTC before, so I didn't know what I was about to see when I turned on the TV. All the pieces from my dreams the night before seemed to come together in this one event. It was horrible.

I am being completely honest here and have not embellished one bit. I really wish this wasn't true. It troubled me so deeply that I couldn't finish the semester at school and failed and dropped out of most of my classes even though it was my senior year and I was getting ready to graduate. I have tried to get over this for years, but I still have horrible guilt and nightmares about it.

49. SYNERGY00

I'm not sure how many of you are familiar with Allegany National Park. Its a park on the border of NY and PA. The road is Wolf Run Road, though on the park maps I believe it shows up as Dead End Road.

One night after it rained my family and I were riding in our van when we stopped on this road so my dad could collect some frogs. as he was standing at the side of the van putting the frogs into a container there was a VERY loud , deep throated growl from behind him. He spun and had me shine the flashlight in the direction of the growl and there was nothing there. At the side of the road there's about 10 ft of grass before the ground rises so steeply it nearly forms a cliff face. After hearing the growl, my dad ran into the van and we hauled ass out of there.

Another time, on the same road, I was walking on my own on one side of the road while my sister and friend were on the other. On the side of the road they were on, there's some logging roads that run a short way into the woods. I noticed they stopped and were just standing there looking at something. When I walked over, thinking they seen an animal or something, I seen a large black human shape down the road just standing there. The shape was all black like it was standing in shadow, except it wasn't. The part of the road it was on was in broad daylight. I ended seeing the exact same shadowy figure a few years later on the same road after hearing very loud crashing noises as it walked through well lit woods.

50. IAMWILLIAM

When I was about 9 my parents had our house fumigated for termites. Since we had to stay out for a few days we got some clothes and rented a room in a motel on the edge of town. It was a two bed room, and my brother and I had one bed and my mom and dad had another.

It was summer, so our parents just dropped us off at our friend's house during the day, then pick us up at night and we'd sleep in the motel. So the first night, right before I fall asleep, the bed shakes a few times. I think its my brother, so I kick him and tell him to knock it off. It happens a few more times and I keep telling my brother to knock it off, and he keeps saying that he isn't doing anything.

So anyways, my parents drop us off to play again the next day, we come back, and its the same thing, every 30 minutes or so the bed shakes. The next morning I complain to my dad about the shaking, and to prove to me that there isn't any monsters or anything he lifts the mattress up off of the box spring, and there's a lady tied up with multiple stab wounds.

According to the coroner, she had died that night.

4

50 Unintentional Quotes From Children That Will Send Shivers Down Your Spine

1. D49A1D852468799CAC08

   "I'm watching you make my sandwich so that when you die I will know how to do it."

2. MAGNUS_MAXIMUS

   My sons were about 2 and 4 when their pet goldfish died. I attempted to use the situation as an opportunity to discuss death and mortality. After I finished my explanation, my four year looked up at me with his big, blue eyes and asked, "Mommy, someday, will you die?" My heart filled with love and a little sadness, knowing this was one of those pivotal moments when the first bit of childhood innocence was lost, and I told him yes, someday, mommy will die.

   "Good," he said with a totally deadpan expression, and walked out of the room.

Later when we were about to flush the fish, he asked if we could eat him instead. I said no, we don't eat pets because we love them, and he said, "When you die, I'm going to eat you."

3. RONEARC

My daughter was only around 18 months old when she uttered her first, full sentence.

She loved to lift the air conditioning grate in the floor of our bedroom and stuff her toys down there. Well, there were some sharp edges in there, so my wife didn't want her sticking her hand in the duct, so she screwed down the edges of the grate, so our daughter couldn't lift it up any longer.

Well, my daughter has this plastic, toy hammer, and she's trying to get it into the vent, but it won't come up. So she takes the hammer, pounds on the grate some, nada. Then she tries to pry the grate up, no go.

At this point, my wife and I kind of chuckled, and our daughter heard us. So she glares with this furious look on her face, throws the hammer at the wall, and almost shouts, "I don't have to take this shit any more."

4. ALFREDFJONES

3-year-old Brother:"If God looks after people, who looks after God?" Mom: "Well, I don't know…" 5 minutes later Brother: "I think the Japanese."

5. ___MADS

I'm the child in this story, but let me tell you about when I was four.

I'm named after my maternal grandmother who died about two years before I was born, and it's worth mentioning that I was the first grandchild born after her death. I was always very curious about her as a young child... one day my mom laid down for a nap and when she woke up, i was standing at her bedside and looking down on her. Apparently I said "Do you remember when I was the mommy, and you were the baby?"

6. WANTTOPLAYBALL

My kid sometimes says creepy things about my husband when he was small. Like, "When Daddy was a baby, we didn't know what to name him, so we settled on Tim." I think she's trolling all of us, but sometimes it is very creepy. My husband's mother died years ago.

7. GODLESSATHEIST

I said something pretty damn creepy to my parents when I was about 10.

So apparently I was making strange noises in my room and my parents both came in to check out what was going on. The moment my parents come in I scream "Ghost, Ghost! Go away" I had a cross necklace so I put it out in front of them and continue to scream "ghost"

Then I scream "Sit down!" and apparently it scared my dad so much that he actually sat down.

The next morning my parents asked me if I was alright. I had no clue what happened and had no recollection of ever saying any of that.

8. MRSANTHROPY

I was putting my daughter to bed one night when she was around two. She said, "Mommy, who's that?" "Who's what?" I asked. "Those people talking to me. In my closet. Who is that?"

I just about shit myself.

9. AMANDAHUGGENKISS

My two-year-old said there is a fairy in his room. He points to the corner with the air conditioner. He says it most nights. One day I was showing him some old family photos. I show him one of my mother and he points to it and says, "Fairy fairy bedroom." The photo was of my mum as a girl. She died 4 years ago.

10. GOOSE_BERRY

When I was 3, I was sleeping in my parents bed when I sat straight up and asked "Mommy who is that man in the corner?" She was terrified. This happened every night until she went to the corner and talked to him asking him to leave us alone because he was scaring me. Still believe in ghosts because of this.

11. NICKDNGR

My mom tells me that when I was a really small child

we would visit my grandfather's house and often spend the night. She says that once, in the middle of the night, she woke up and I wasn't in the bed (young enough to co-bed). She got up and I was standing in the living room with my hand in the air like I was holding someone's hand and I said something along the lines of "not being able to go with you because my mom didn't say I could." We didn't spend the night at my grandfather's house again for another decade.

12. WHISKEY_SOUR

Once I was taking a nap on the couch. I was waking up, and just as I'm opening my eyes, I see my 2yo son walking toward me with a serious look on his face. He leans in close and whispers, "It Happened." He then leaves without another word.

FOR THE LOVE OF ALL THAT IS HOLY, WHAT HAPPENED??

13. SECOND_LOCATION

My kid was in the bathtub one night with the bathroom door open and I was puttering around in the next room. She called out and said "hey mommy, who was that blue guy who just walked down the hall?" She said he was tall and thin and featureless like "the shape of those men on the bathroom door like at a restaurant". Creeped me out!

14. PJDAVIS

My son (6), when asked what he wanted to do when he grew up, said, "I think I want to be a fighter pilot, or maybe a funeral worker like daddy."
I am a software engineer.

15. GENERALOFFENSIVEUNIT

Getting my two and a half year old daughter out of the bath one night, my wife and I were briefing her on how important it was she kept her privates clean. She casually replied "Oh, nobody 'scroofs' me there. They tried one night. They kicked the door in and tried but I fought back. I died and now I'm here." She said this like it was nothing. My wife and I were catatonic.

16. CYPRAH

My little brother said something eery to my grandma.
"I like this mummy better than my last mummy. My last mummy locked me in a room and I drank some paint and died."

17. ETWAS_NAHT

The rare occasions in which small children have alluded to having violent experiences that led to previous deaths freak me the fuck out.
The most detailed one I ever heard was actually delivered second-hand through my friend's mother. Apparently beginning around the time my friend could form sentences until he was little more than 2, he would go on and on about how he was a Native

American named Conchon and that after his wife and son got sick and died, he moved to a mountain to live by himself with his horse. He died of a broken neck when he fell into a ravine. Weird shit, man.

18. POLITE_WEREWOLF

When my father was a kid in the 60s, he would go into the living room in the middle of the night, turn the TV on to static, climb on his rocking horse and slowly rock back and forth in the dark room only lit by the TV static and slowly say "I hate mommy. I hate mommy. I hate mommy." over and over again. My grandmother says it was the creepiest thing she's ever seen. Well, that and the UFO.

19. ALPHAREDDITOR

My godson told me that he was "fully erect and ready to wreck." He was 3. His dad told him to tell me that and is a twisted man.

20. BLT2002

My one son was eating chicken nuggets and he would always eat the breading off it first, he takes a bite of the breading and then says "Oh no! Your face is missing!"

21. KAYAXIALI

My 3 year old was laying on my chest a few weeks ago and she said "I can hear your heart, Mommy. It was much louder when I was inside there with the poops that didn't come out yet".

22. GEEKAMONGUS

My son was two. He was in a pattern of waking us up at about 5:00 am every morning. One morning I took him downstairs and plopped him in front of the TV so I could try to go back to sleep for about 30 minutes on the couch (right by him).

I woke up a few minutes later and he was standing in the foyer, pointing into the kitchen, laughing. He then said, "Mommy is floating in the kitchen."

I didn't think much of it…went back to sleep for a bit.

About 30 minutes later his mom came downstairs having just woken up, saying she "had one of those weird dreams where she flew out of her body, went downstairs and found herself in the kitchen."

Freekay.

23. PHANTOMSERIOUSLY

I shared a room with my older sister and we had huge closets in our bedroom that were about 6 foot tall. My mother would wake up in the middle of the night to hear me crying and she'd come in to investigate what was wrong. She then would find me sitting on top of the 6' closest, cross-legged and rocking back and forth while crying about; "The big scary man put me up here". Since my mother was tired from it being the middle of the night and being heavily pregnant she didn't really think about HOW I got up on the closet, but would put me back into bed and comfort me until I fell asleep again. But then my grandmother came to stay with us a few nights and she told my mother that she woke up in the middle of the night because it got suddenly cold and her bedroom door handle was turning. The door opened but no one was there and then the bathroom door opposite her door opened on its own. She stared out the door for a few minutes not moving because she was in shock and frightened, but then heard me start crying. My mother walked by her room to get to me and of course I was crying about the man putting me up there. My grandmother told my mum what she experienced and my sister slept with my Gran and I slept with mum for the next couple of weeks after that. It stopped once my brother was born, and to this day I have no idea what really happened.

24. ELK_ATTACK

I had a music teacher, who took his 4-year-old daughter to an old theatre in Alaska. She started crying immediately when she walked in, so he took her outside – and she stopped crying. He took her back in, she started crying again, so he took her outside again. He asked why she was crying, and she said: "That's where the people with no eyes watch you."

25. JIMPARSONSROX

My dad watched his mother die of a ruptured gal bladder when he was twelve and still remembers vividly.

My sister, one day, randomly gets up almost an hour after she's gone to bed and goes up to him. The conversation went like this:

Sister: Daddy, your mommy died in a red sweater, jeans, sneakers and with her hair in a ponytail, right? And her hair was blonde?

Dad: Drops book he's reading and stares, wide-eyed, and then says Yes...

Sister: What color were her eyes?

Dad: Blue... why?

Sister: Oh, she doesn't have them anymore, just empty sockets. I was curious.

And she goes right back to bed.

26. YGMIC

When I was probably around 6 or 7, I had no idea what sex was or how making babies worked. I just thought that if you loved someone, you'd eventually end up being pregnant and having a baby. So one day, I randomly said to my dad "one day I might end up having a baby with you". Which probably creeped him out no end.

27. PUPPYBREATH

My cousin's kid when he was around 4 or 5 came into the bathroom as I was straightening my hair. He closed the door, looked at me and said: "I don't want to kill you."

Creepy. He's 13 now and whenever I tell him the story he just laughs his ass off.

28. [DELETED]

My 6 year old daughter came downstairs from her bedroom and said "Dad, i think Kacey is dead", that's my 3 year old daughter. Of course i ran up to make sure Kacey was ok, at which time my oldest daughter raided the cookie jar.

29. ASSNECK666

My son when he was about 2... he had a weird fear of being abandoned, which there never was an incident of him getting lost or any type thing. He asked my wife if we have ever forgotten him anywhere, which she replied no. He responds "oh that's right, it happened when you were small and I was big"

30. SUPERFISH1984

My daughter had some imaginary friends for a couple years named Dodo, Ghana & The Evil. They just sort of appeared out of nowhere when she was about 2.5 years old. It started with Dodo and Ghana, then a few months later (she was about 3 at this point) she came up to me and told me with a creepily expressionless face: "The Evil is coming over today" and just walked away.

Turns out, The Evil was actually a pretty nice imaginary friend, she just had an unfortunate name.

31. NICKDNGR

I was in Fred Meyers buying groceries a few months ago and a random kid walked up right next to me and pulled on my pocket. Kid had to be like 7 or 8. Anyway, he pulls on my pocket and I look down and he makes a throat slicing motion with his finger over his throat and just walks away. That was the creepiest run-in with a stranger, not to mention a kid, I've ever had.

32. PANSHAKER

I was changing my 2 1/2 year old daughter's diaper when she reached up and touched the side of my face. She looked in my eyes, and said, "I love you, but I never should have married you." It was a week later that I realized the babysitter had showed her "The Fantastic Mr. Fox", and that it was a line from the movie, not something my wife was practicing saying in the mirror.

33. CAEHTRULY

I'm not a parent, but my mom told me that when I was really young, I used to sit by the rocking chair and mumble things. She eventually became curious and asked what I was saying. I told her I was talking to "the old lady".

Also, about 10 years later, I stayed at my aunt's house. In the morning, I heard my 3 year old cousin stirring, so I decided to be a thoughtful niece and went to get her out of bed so my aunt could sleep in longer. When I walked into her room, she stood up in her crib and said "Your friend came and woke me up last night." I was staying in a room by myself the night before.

34. ICYCOOL

I've got a few stories from my own childhood actually.

I remember one time my mom told me when I was about 2~3 I told her that I was once a firefighter and died fighting a house fire.

There was another time when I was 2 years old, at my grandmother's house, when I inexplicably walked up to the glass coffee table and smashed my head straight through it. I didn't cry or say anything, just did it. I still don't know why or even remember doing it.

35. MAIMER__

Little brother was talking to someone in the living room while i was fixing lunch, i go into the other room and ask him who he's talking to. Hes looking at the window and says this old lady, she's very funny but doesn't like you.

36. MOKAFUZZ

Not a parent but I've worked at a daycare centre. I was watching a little girl playing with one of the dolls. She was dressing it, putting it to bed, etc. At one point she put the baby on the toy stove. I asked her what she was doing, "I'm burning the baby," she replied.

37. PAULA36

Well when I was a kid I slept walked one night and it freaked the shit out of my dad.

My dad heard a strange noise in the front of our house and walked out to see me sitting on our front step with the door open in the middle of the night. He asked me what I was doing, and I turned around and said, "I'm waiting for someone".

I had no recollection of it in the morning. He was creeped out for quite some time.

38. GAVINARDO

My parents were asleep. I apparently began sleep walking, never done it before, and never since that I know of. I went downstairs – I was about 7 I guess – and began rummaging through the kitchen. My dad apparently woke up soon after, and heard the noises. He was alarmed, and woke my mom. Told her to go make sure I was okay. He quickly got up, fearing an intruder in the house, and grabbed his .38 revolver. He hustled downstairs, held the gun up and flipped the lights on, only to see me having taken everything out of the fridge, arranging it on the floor. My mom discovered I wasn't in my bed about this time I guess, and screamed for my dad. Her scream woke me up, and I was startled to see I was in the kitchen, food all around me, and my dad very frightened and holding a gun. My mom put me back to bed, and I woke the next day thinking it was just a strange dream.

Years later, as an adult, I realized it may have been real. I asked my dad, though he cringed while discussing it. Understandable. He pointed a gun at his own son.

Very scary night for him, more than me.

39. [DELETED]

I used to have night terrors when I was around 2. Vivid nightmares that involved walking and talking in my sleep. Consequently, I often spent the night in my parents bed. One time my mom woke up and saw that I was missing. She found me standing in the living room. She tried to pick me up but I backed away and screamed, "Wash the blood off your hands!"

Said it creeped her the fuck out.

40. KIMDEALZ

My son was 3 when I was tucking him into bed one night and he said, "Mommy what's that big thing?" I replied, "What thing, baby?" THEN HE SAID "That big thing right behind you"

I knew there was nothing right behind me but a wall, I just did the scumbag mom thing and backed out of the room and shut the door. He wasn't scared of 'that big thing' but I sure as hell was… creepy kids!

41. KELLEYSARAH7

I was the kid, my mom told me this story once I was older. My great grandfather died, and because I was so young no one told me. My mom took me to his grave a few weeks after it happened, and let me play amongst the gravestones while she lay flowers. As we were leaving, I stopped and asked "why is great grandpa sitting in the tree?" I then pointed to what appeared to my mom as an empty tree, and waved. The tree was planted so the branches hung right above where he was buried. TL; DR: Pointed out my great grandfather to my mother without knowing he had died.

42. CHALLENGEREALITY

Not my kid, but the kid I was babysitting. And perhaps more sad than frightening.

I used to babysit two brothers, one was 9, the other 4. The 4-year-old was a pretty typical kid, while the 9-year-old was really distant and sometimes downright cruel. He would flip out at his younger brother (physically and verbally) for the smallest things and would laugh if the younger hurt himself, etc. It was tough mediating between them, and the parents seemed oblivious to how much the older brother loathed the younger. I figured the age difference was really all it was, but sometimes I sensed a real hatred radiating from the older brother.

So one day at the playground the 4-year-old's friend comes up and says the younger brother has been telling everyone, 'My brother is a killer!'

I pull the 4-year-old aside and say, 'Hey, it isn't nice to tell your friends that your older brother is a killer.'

To which he dispassionately responds, 'But he is a killer. He kills me every day.' It was like the younger had given up. Really disconcerting to see a 4-year-old so.... I dunno, hopeless.

43. FIDGETSTIC

My ex and I took our kids over to my dad's house and we were up stairs in one of the empty rooms playing. The closet door opened a little bit on its own and my three year old jumped up, ran over to it and said "it's grandma". My mother died before he was even conceived.

44. UNKAMENRIDER

When my nephew was 3 or 4, he would stare at the window in my parents' kitchen. One day, my mom asked him what he was looking at, and he said, "When I lived here before, my name was Alphonse, and I was bigger than you." My mom was slightly creeped out and eventually told my stepdad. My stepdad just kind of blinked and said, "Hmm. That was my grandfather's name, but we don't talk about him."

45. THEREVENTON

A story of when my brother was younger that my parents often tell.

My dad and my granddad went to an old army base that was now used as a museum. As they went into a bunker, my brother started saying "This is where we hid from the monsters that went 'BOOM BOOM BOOM'". My dad thought it was a bit strange but kept on looking around. A low flying plane flew by and my brother said "That is the monster there." as he was pointing to the plane. My brother was 2 at the time, which must have been in 1994. Needless to say, both my dad and my grandad were creeped out by it, and still are today.

46. SDAVIS213

While Playing classroom with my three-year-old brother he made an off handed comment about being in my mother's belly twice. I was amused and said oh really. He proceeded to tell me in amazing detail being inside our mother. He told me about it being warm and that he liked it but he always felt sick. One day he got so sick that "they" came and told him he had to leave. He didn't want to so they promised him he would get to came back again and back to our mom. So he left and they let him come back again and this time he didn't feel sick. I lost my mind and started screaming for my mother. He told her the same story then after she stopped crying we were not allowed to talk about it. I was ten and I was old enough to remember that she had a miscarriage almost 1 year before my brother was born.. Side note i'm not religious but my brother always kinda makes me wonder.

47. FLOWERSCANDRINK

My daughter was 4 years old. One morning I heard her door open and shut. That usually meant that she would be coming to our room to lay down with us. She never came in, but shortly after I heard her voice. Hoping she would go back to sleep I let her be for a bit. Then I heard the door open and shut again. This time I decided to go into her room and see why she kept getting out of bed.

I walked in and she had her eyes closed.

"Sweetie?"

"Yes, daddy."

"Why did you get out of bed?"

"I didn't, I was trying to sleep but he wouldn't leave me alone. He kept talking to me and asking me questions."

"He? Who is he?"

"The little boy that was in my room."

"Umm, sweetie that was just a dream. There is no boy in your room."

"I know that. He just left."

"Ok, well what was the little buy doing?"

"He was hanging from the fan and asking me a bunch of questions."

"How was he hanging from the fan? With his arms?"

"No, with a rope."

Scariest fucking moment of my life. I asked her about it about a year later and she said she doesn't remember.

48. DRUNK_ELECTRIC_FIRE

When I was little my mother took me to a petting zoo. They let me milk the cow and everything!

Apparently upon getting home, I yanked my pants and underwear off and asked my mother to milk me.

49. [DELETED]

When I was pregnant with my first, a five year old came up to me and said, "All babies are born alone." He mom and I share freaked out glances and then she awkwardly tried to fix the situation by talking about twins.

My oldest (6, 5 at the time) once got mad at my youngest (3, 2 at the time) for sitting in the doorway to their room. Instead of asking him to move or calling for me, my eldest grabbed the little one by his head and shook him as hard as he could. I freaked out, scooped up the toddler and yelled: "What's the matter with you?! You can't shake babies! Do you know that could kill him?" Without flinching, my oldest looked me in the eyes and said, "Yes, that's what I was trying to do." I lost my mind and called up his therapist, wondering if I needed to or could commit him. It was really scary.

To clarify, he had a tendency to react physically, but before that I never thought he was actually trying to hurt anyone.

50. SOPA_NO

When my oldest brother was 3 or 4, he fell into my cousins' pool. The pool didn't have a ladder and was several inches to reach to get out, no way for a 3 year-old that has ever swam in his life to escape. Well, while all the adults were talking they hear him screaming and splashing, and then silence. When they ran to get him, he was standing right next to the pool and he was soaking wet. When they asked him what happened, he said, "I fell in the pool and couldn't get out, then a shiny man pulled me out."

That one will always give me shivers.

5

50 People On 'The Time I Met A Celeb And They Acted Like A Total Weirdo'

1. j_patrick_12:

    I just went through the LAX security line with Marilyn Manson. He had "FUCK" scrawled in large letters across the bottom half of his face, with what appeared to be a grease pencil. As we each removed our boots in the security line, he kindly explained that it was not directed at me or anyone else in the airport, but rather at the paparazzi, so that they couldn't sell any photos of him that they took. He was really apologetic about it, and covered his mouth around young children while apologizing to their parents for exposing their child to profanity.

2. Accidental_Feltcher:

    A buddy of mine was having a smoke outside a bar near Detroit, when Bruce Campbell strolls by. It's a little after 1 am and he's wearing ray ban shades. One of his friends sheepishly asks "Excuse me, are you Bruce Campbell?". Bruce stops, tips his sunglasses, and responds with "Well, someone's gotta be".

3. silentlamp:

Served Christian Bale a latte at work, he went and stood facing the wall, like a child's punishment until his coffee was ready. People were more bemused by his behaviour, thus giving him slightly more attention.

4. strangeclouds:

When I was about 10 I was in Chef Mickey's at Disney world, I had about 20 strips of bacon on my plate when a voice from behind me say "easy on the bacon young man" and there was Arnold Schwarzenegger. I told him he was the terminator and he laughed and signed my arm.

5. ALL_CAPS:

Lou Diamond Phillips tried to buy weed from me at a pool hall. One of the few times I regretted not being a drug dealer.

6. PatBabyParty:

I was having lunch at a cafe in Culver City with a friend when Nick Swardson walked by our table. I said "hey Nick! Can I get a high five? I loved you in

Grandma's Boy!!" So, he gave me a high five and then asked me if there was room for another at our table.

We were both somewhat confused but moved over and made room for him, so he sat down next to us and took a cookie wrapped in cellophane out of his pocket and said "Hey, do you guys want some of this cookie? I just got it at the counter, it's so fucking good!!" and proceeded to break us both off a piece. He asked how our day was going and if we were enjoying our food, then said "it was great meeting you guys, I'm going to go get really drunk now! Take care and keep being fucking awesome!" and walked off.

7. catangel001:

One time I was in a Subway (sandwich place) in Orem, UT and Gerard Bultler walks up behind me. I look at him, he smiles at me, and then we both pretend that he isn't himself. I say, "You look like Gerard Butler, he's one of my favorite actors." He said, "I get that a lot," and winks. Then he asks, "Well, are you a true fan of his?" And I say, "Of course!" He asks if I knew what he used to do, I reply with, "He used to be a lawyer, before giving that up to pursue acting," and then we spent the next twenty minutes discussing law, politics, and why someone would give up a successful career in law. Then, we shook hands and parted ways.

8. whomwhom:

As a child, I remember that Bill Cosby and Dianna

Ross both came to host some corporate event at a large tourist hotel in my beautiful small town.

Dianna ross evidently got pretty drunk, and didn't make the event, and it was cancelled. It was big news the afternoon it went down, and some of my friends and I were hanging out at the skate park down the street from the hotel.

A Rolls pulls up, the back window rolls down, and Bill Cosby sticks his head out.

"HEY! Have you kids seen Dianna Ross?"

"...no."

"She's missing."

"...oh."

"How are you kids doing, then?"

"...good"

"Ok. Have a good day, watch out for Dianna Ross."

"...thanks"

drives off

9. dadamax:

I used to live in the East Village about a block from Willem Dafoe. I would see him around the neighborhood a lot, enough times that we would nod to one another in greeting as we passed on the sidewalk. One day I walked into our corner convenience store and I completely spaced about why I came in there. I stood just inside the door trying to remember what I came for when I hear the bell on the door jingle and I turn around and see Willem Dafoe standing behind me. It was a small store and he thought I was standing in line at the counter so I politely told him to go ahead of me because I have no idea what I needed. He steps in front of me, stops, and says, "Dammit, now I can't remember either." After a few seconds he snaps his fingers, reaches up on shelf and pulls down about five packs of condoms and giddily throws them on the clerk's counter. I told him I just remembered that I only came in for some dish washing liquid, got it from another shelf and stood behind him to pay. After he pays, on his way out the door he turns around to me and says "It's gonna be a big night!"

10. hoodoo-operator:

My mom met Frank Zappa, and he ate a salad with his hands.

11. WookieProdigy:

I saw Hulk Hogan at an Apple Store in Tampa. It was after his divorce and he was with his new girlfriend. I was working up the courage to go and ask him for a photo with him; Hogan is an intimidating dude. When I finally resolved to approach him, he started to cry. At least I think he was crying; he was hunched over with his head in his hands and his shoulders were moving up and down as if he were sobbing all while his girlfriend rubbed his back.

    I decided it probably wasn't a great time for a photo.

12. way_worth_a:

I was a patient in a behavioral health care facility,

doing a puzzle when I heard the patients say, "That's Steven Tyler." He was getting treatments for painkiller addiction at the time. Decked out leather jacket, huge mouth, hair, entourage, everything.

He was about at the lobby ready to leave, so I had to think fast. I had my acoustic guitar while I was in the hospital, so I asked him, playing dumb, "Are you a musician? You look like one." He said, "Yes, I am." I asked him, "Would you like to play my guitar?" He looked at what might have been his agent, who gestured at her watch, but said, "I have 5 minutes, alright."

We sat down in the lobby where the patients got their vitals, I got my guitar from the nurse's office, and he proceeded to tune my guitar to something open and ridiculous like DADAAD. He sang a new song about rain, purple rain, and washing things. Since I beatboxed, I laid down some drums while he played. Afterwards, he said to me kindly, "That's the first time I've played music sober in two years. Thank you."

13. robotrock1382:

Years ago, I'm with some friends at this shitty bar in New Orleans. A friend of a friend is playing an acoustic show, and we're the only ones in the bar. Out of nowhere, this giant crowd comes into the bar, and out of nowhere, Nic Cage emerges. Where we're sitting, between us and the stage, is a dance floor. He falls to his knees, and starts doing this weird dance thing. It looked like the pic of Hendrix when he lit the guitar on fire. He does this for a very short amount of time, then he hops up, goes " Woooohooo" and saunters out the bar, quickly followed by all of his followers. It was fucking surreal.

14. CalicoBlue:

I met Bill Clinton when he was doing book signing for My Life. I told him he was the sexiest president we've ever had and he shook my hand twice. My friend did not get a second handshake.

15. jd1z:

I met Alan Rickman as he was going into the stage door of the theater for the broadway show Seminar. He was walking past me and I quickly blurted out, "I think you're awesome!" He stopped, turned slowly towards me, extended a hand to shake mine, raised one eyebrow, and said,
"Likewise."
It was awesome.

16. healthyjorts:

my friend John was playing golf with Arnold Schwarzenegger (my friend is Arnold's lawyer's son). Arnold was about to tee off. Arnold lines up for his drive and says, "So... John, when was your first blow job?" My friend nervously replies, "uh... seventeen" to which Arnold grins, winds up for his shot, and ask, "how did it taste?" He then took a massive cut at the ball only to shank it to the right.

17. watchthatcorkscrew:

Not technically me, but got into a cab and the driver immediately started telling me his last fare had been Ralph Fiennes. Apparently RF noticed the pictures of his son on the dash and driver admitted his son was a massive fan of Harry Potter, so Ralph offers to give him a call. So the guy calls his son and says 'I've got someone very special here who wants to talk to you', Ralph takes the phone and goes straight in, full Voldemort voice, with 'So I hear you think you're a strong enough wizard to defeat me??' Apparently for the next few minutes all the cabby can hear is lots of tough talk and then a lot of shouting 'expelliarmus!' 'you'll never defeat me!' and then a very convincing death gargle. Ralph passes the phone back, signs a bit of paper which the cabby showed us 'To George, The greatest wizard I have ever duelled' and then tips very well ? pretty fucking cool I thought... Not a lot of kids get to duel Voldemort over the phone...

18. bluejams:

My dad got stuck on the median of Park Ave with John Lennon who was crossing the other way. My dad said "Surprised to see you here" and John answered, "well I'm surprised to see YOU here". Then they just walked there separate ways.

19. megasaurasrex:

Sylvester Stallone came into my Urban Outfitters and

bought this coffee mug we have where the handle is brass knuckles. Actually, he bought every single one we had.

20. shazbotabf:

I was in Long Island visiting my friend, and we had just woken up from a long night of drinking. Feeling the way that we did, we decided to go to a local deli to get some breakfast sandwiches, because that shit is the bomb. We waited in line for a while, and when the person in front of us was ordering, in walks muthafuckin BILLY JOEL. Now i'm a pretty big fan, so I have a silent freakout in my head, wondering if i should ask for an autograph or whatever. So Billy Joel grabs a gatorade, and then walks to the front of the deli line. Like, in front of me. So, the other guy is still ordering and I think that like maybe Billy Joel just wanted to look at the menu behind the counter or whatever. The guys finished ordering and pays, and then Billy Joel just fucking starts ordering like a million damn sandwiches. At this point, my irkedness at him cutting me overtakes my starstruckedness and I say "Oh hey, I think i was next man." He turns around with the most contemptuous look on his face and says to me like I'm a piece of shit (I'll never fucking forget this) "Well I guess now I'm next, you shit." I just stood there with my mouth open. I didn't know ANYONE could be that much of an asshole, let alone Billy Joel. FUCK that guy.

21. Buckfutters:

I once had to throw Pauly Shore out of a strip club because he was snorting cocaine off of his table.

22. breakinthehymen:

I was behind Keanu Reeves at a local coffe/smoothie shop in Santa Monica, called Manny's IIRC. When he ordered his smoothie he asked, in his best Bill and Ted voice, "Can I, um, like… Have a smoothie?"

23. JewChooTrain89:

I worked at Morgan Freeman's production company as an intern three summers ago. One day I was instructed to bring a bunch of old film cans to a storage unit. I was carrying about five or six, since I didn't want to make two trips to the car. I get into the elevator and take it down to the garage and when the door opens all I hear is "BOO!!" I dropped all of the film cans onto the ground and one of the cans opened and the filmed spilt everywhere. I look up and Morgan Freeman is standing above me. All he says is, "Looks like you have some work to do." He then enters the elevator whistling and closes the door. Right before the door closes, he winks.

24. newo32:

A few weeks ago, I worked for author & pick-up artist Neil Strauss. It was actually my last day on the job, even though I didn't know it.

I was up in Malibu, waiting for the bus to take me

down to Santa Monica, when a car pulls over, and the driver goes "I'm going to Santa Monica if you want a ride!"

"You're not gonna kill me, are you?"

"No, I'm not gonna kill you."

I get in the car and he extends his hand and asks me my name in the MOST recognizable voice ever.

I look up...it's MARTIN FUCKING SHEEN. We shot the shit for like 45 minutes as he gave me a ride down the coast, and INSISTED upon taking me to my specific bus stop.

At the end of the ride, we chatted really briefly about Charlie (after talking about Sorkin and West Wing and me just sporting a general writing boner)...and then he told me to hang on a second.

He gave me a blessed Rosary from Jerusalem. Every bead was made from an Olive pit. He asked me (a "recovering Catholic") to say a Hail Mary for him, and one for Charlie.

I said like six.

25. the_fewer_desires:

I spent an evening with Elijah Wood in a hot tub. We argued about smoking laws. He made out with a girl who thought he was Toby McGuire but kept calling him Toby Keith. We hugged before he left an our bare chests touched.

26. xeb_dex:

Saw Seth Green in line checking out at Toy Tokyo

in NYC during comic con last year. He was buying THOUSANDS of dollars worth of toys while his girlfriend (lady friend? female acquaintance?) was standing near by, arms folded, tapping foot annoyed as FUCK.

27. gyakutai:

Not me, but I used to know a driver who would pick up Christopher Walken from time to time. He said Mr. Walken was one of his favorite customers because he'd sit in the front seat, tip well, and was an all around great person. So the driver, Peter, picks up Mr. Walken and is driving him around and he mentions that his friend is a huge fan of his acting. Being the good guy he is, Mr. Walken says "bring her next time you pick me up." Fast forward to the next pick up, Peter's friend is in the front seat and has no idea why she is there. Peter gets Mr. Walken from the arrivals terminal and comes back, while Peter grabs her attention, the actor slips into the back of the car unnoticed. After Peter starts driving, Mr. Walken says hello which startles the girl. She turns around, sees Mr. Walken and faints in the front seat. A few minutes later, she wakes up, gets an autographed photo and enjoys the ride home knowing she passed out in front of one of her favorite actors.

28. Disco_Panda:

In the Westin Bonaventure in downtown LA (you've seen it in half a dozen movies). One of the elevator doors open, and everyone rushes out not making eye contact with anyone, all pale and disturbed by something.

Moments later, Christopher Walken strolls calmly out of the elevator eating popcorn.

No idea what happened in that elevator

29. whereitsbeautiful:

One of the guys from Less Than Jake once spent about 15 minutes petting my hair while he signed autographs for other people.

30. ShavingFoam:

I was an extra in the dark knight rises.

I was an orphan in the orphanage, when filming, some children where being rowdy and Joseph Gordon Levitt told them that this was "dead fucking serious." The kids where terrified by that and shut up after that. I understand why he did it though, nobody wants to watch rowdy kids in a film anyways.

31. Waywardcross:

I worked at Blockbuster video a few years ago, and I got to tell Al Gore that he owed us 1.50 because he returned "I Love You Man" late.

32. sneekymoose:

I met Carlos Santana on a beach in Maui while on vacation with my family. He came over, told me I looked upset, and that I should enjoy my beautiful family because not many are as blessed as I am to have a family so beautiful. I thanked him for his advice, and told him I loved my family very much. He asked my name, we shook hands and parted ways. Always comes to mind when I'm sitting around enjoying my family.

33. eraspells:

I was staying at my friends house in Windsor when I was 9 and she happened to live nextdoor to Anna Friel and David Thewlis (when they were still going out in 2003).

Just so happened that Prisoner of Azkaban (this is important) was being filmed at this point in time, and my friend's house shared a communal garden with Friel and Thewlis (Professor Lupin, if anyone doesn't know).

It was a sunny day so most people were sitting outside having a picnic in this massive garden, after we'd finished eating us kids started having a water fight against the kids who lived in the other house, and eventually I think quite a few adults joined in as well.

Suddenly I am hit full on in the face by one of those super-soaker things, I look over at the perpetrator and, to my surprise, it is Gary Oldman. It was so cool it almost held me back from defeating him.

Oh and later Thewlis attacked me with a hose.

34. [deleted]:

A few years ago i was at a blockbuster in Bend, Oregon, checking out Evan Almighty, when i hear a godlike voice say "The first one is better". I turn around and Morgan Freeman is right behind me, smiles, winks and walks out the door.

35. cuntpunter2001:

Was dating a comedian a few years ago so I went to quite a few shows. One night he brought me back to the greenroom to meet Tommy Davidson (who's just about the nicest guy.) I walk in and sitting on the couch is none other than OJ Simpson. He promptly pulls me down onto his lap and has his drunk girlfiend, who is the spitting image of Nicole Brown Simpson, take a picture of us. She then joked about how I shouldn't be afraid for my life because "he doesn't murder brunettes".

36. batsam:

Bill Murray showed up to one of my college football games last year. The marching band noticed and played the Ghostbusters theme song, and he got up on a podium and started conducting them and dancing around. Nobody knows why he was even there to begin with. Our football team sucks.

37. beepbopborp:

My pal was dating Danielle Fishel (Topanga) at the time and brought her to our company Christmas party. I'd met her a few times before that when he brought her around, but while we were catching up, she said to me, "I've been taking pole dancing class! Want to see?" This is where she proceeded to do a "nice" little dance on an awning pole for me. It was very awkward, yet extremely satisfying.

38. Charawr:

When I was four I was at a baseball game with my parents and their friends. My sister and I were enjoying some soft serve ice cream cones, when out of nowhere someone is trying to eat my ice cream! Everyone is laughing and telling me it's ok and to share it, but I was not having it and refuse to let the man have a bite. I later found out that man was Bill Murray. There are pictures, but my parents will have to dig them out.

39. shrkdvr:

I was once assigned to shoot video of Paul Simon for my employer who had hired him to play a show. He refused to speak directly to myself or my crew so standing three feet away from him I had to direct my questions to his publicist who would then repeat my questions back to him. He would then direct his responses to his publicist and she'd then repeat his answer to me. It was completely insane. That night I threw away all of my Paul Simon albums & haven't listened to him since.

40. fire_eyez:

Tony Geiss (voice of big bird, Oscar the grouch and most of sesame street) was a member of a nudist colony I visited.. I said to my friend "I'm going to kill you" and and Tony chimed in with "don't be killing people(Oscar voice)" – "it will ruin your day!" (big bird voice)

41. Wiro8743:

Was seeing Billy Elliott in the Victoria Theatre in London circa 2005 with my uncle, when about 15 minutes into the performance, Michael Jackson comes in surrounded by 5 or 6 bodyguards. They came and sat behind us, and as one of the bodyguards was walking behind us, he tripped and dropped the programme he was carrying, and it fell on my lap. I turned around and gave it back to hear from the King of Pop: 'Sorry young man, fuckwit here can't stand still'. Best day ever.

42. stevyjohny:

my dad said he met MC Hammer when he worked at sbarros. Went my dad went to give MC his pizza, MC reached out for the plate. Then, my dad quickly pulled the plate away and said "Can't touch this". Apparently, MC laughed it off. I still tell my dad I don't believe its true.

43. 127001y:

I once played golf with O.J. Simpson and Lawrence

Taylor. So, this took place about one month after O.J. was acquitted for the double murder of his wife and Ron Goldman. I was a walk-on at a local golf course in L.A. and was called to the first tee by the starter to join a threesome. O.J.,L.T. and an attorney friend of theirs. O.J. was playing very poorly and getting frustrated on the front nine. After the turn, on the eleventh tee, O.J. hits a shot right down the middle. He looked over at L.T. and screamed at the top of his lungs.. "Nobody Fucks With My Pussy"!!! He said it twice and loud enough for people on other fairways to hear. I was totally blown away. I could see the absolute rage in his eyes as he was screaming. Now, before this happened, I kept thinking about how this guy killed two people and here L.T. is hanging out with him. But when he screamed, I could see it all. No doubt in my mind… This guy is a fucking murderer…

44. valtastic:

A few years ago, I used to meet a friend at Wendy's in Union Sq (NYC) after she got out of class. I'd get Wendy's and she'd get Taco Bell next door and we'd hang in Wendy's eating it. One night, Tracy Morgan came in with some lady. He had just bought her Taco Bell and was getting himself Wendy's. I'd never felt such a camaraderie. He was also acting weird and wearing a wet suit as a shirt and giant cargo shirts and overall just looked like a giant homeless toddler, and I had the revelation that Tracy Morgan = Tracy Jordan.

45. peasoupanderson:

Went to a Jack Johnson concert at the Hollywood Bowl years back. Bunch of friends sitting at the top row of the top section. My friend managed to get some dude to pass down a fake cigarette pipe to him. Within seconds, a security guard nabbed him and was ready to throw him out. Most of us, including myself, were pretty confused because we didn't see the exchange so we too thought it was a cigarette.

After a minute of confusion and deliberation, some lady intervened placing her hand on the security guards hand. "This is a rock concert, these things are normal" explained Julia Roberts. The security guard was puzzled at first, then backed off. My friend got to stay, concert was great, the end.

46. FartyNapkins:

A lady I used to work with ran a boat rental business on Lake Coeur d'Alene and Robin Williams came with two other people and rented a boat from her. Robin stayed on the lake too long and his boat ran out of gas so he called her to come and get him and give him a tow. She said the other two guests got into her boat but Robin Williams stayed back on the boat that had run out of fuel and pretended to be driving it. She said he stood at the helm for 45 minutes having a conversation/yelling/laughing to himself while he steered the wheel back and forth and pretended to be at sea. She also said he was the nicest, happiest, and most energetic person she had ever met.

47. snezel:

My little cousin was walking onto her plane at LAX with her family, as they were going through the first class section they had to stop. All of a sudden, Richard Simmons turns around and says to my 8 year old cousin, "I just love your outfit!" (In his voice if you can imagine it).

48. pineapplehaze:

I was in an elevator with Pete Sampras and he stood in the corner facing the wall until we got to the lobby.

49. dobiscuits:

A friend of mine always tells the story of his encounter with Bill Murray. Whilst walking home from a near-by Dominoes pizza back to the house, Bill Murray comes up behind my friend, takes the pizza and runs away with it shouting "No one's going to believe you!"

50. rainbowbelly:

I met Kanye West at what used to be the Virgin Megastore on Michigan Ave. He was buying Jay-Z's Black Album. When my friend and I asked him why he was buying a CD he had a part in.. He replied with 'I gotta support myself too' proceeded to shake our hands and talk about his work with Kweli. Cool guy despite his arrogant ways.

6

66 Soul Punching, Evil Things People Have Done, Said, And Experienced

People aren't perfect, and this is a perfect example of how bad our actions, words and experiences can be. As I read these, I relived all of my embarrassing and awful moments in life and I just want to crawl into my bed and play Animal Crossing to forget it all. Help me, KK Slider. Help me. Found on r/AskReddit.

---

1. ROBBIESTEW

Back in 2002, I was in a car accident and was in the hospital for some time. My gf ended up cheating on me while I was in there with some Cook from a restaurant who dealt drugs on the side.

Screw her, I was glad she was gone, who wants someone like that in their life. I was beginning to move on, but this asshole would call me occasionally bragging that he took her from under me. I didn't get it. I started to get more mad at him then her.

I found out his email address and was able to get his secret question with some research.

I got into his email and found a picture of him and a good buddy I guess with their arms around each other posing for a picture. Since I was already on bedrest due to the car accident, I took my time did some research digging through his old emails about this friend and other life stuff.

I then spent the day writing a very long and detailed message about him being a homosexual, and how he was coming out of the closet, and he was madly in love with said friend from picture and they were going to get married and try to adopt children etc.

It was really well written and very detailed.

I then send the email out to everyone in his address book and everyone that had ever sent him an email. Golden I thought.

..But it got even better.

Not only was his friend not gay, but he was engaged, and the fiancee did not like this at all, I guess it caused quite a shit storm, because the fiancee thought these guys spent too much time together anyways.

..but it got even better, (or worse).

Apparently the dickhead's mother about 10 years ago, gave herself to god and became a nun. So you can imagine her reaction when her little angel, was doing this horrible sin. (I made references to them being lovers for years in the letter.)

Needless to say, it caused a giant shit storm with everyone in his life.

It was pretty evil, but the guy was an asshole loser, and I enjoyed every minute of it.

tl;dr – Guy was a dick who wouldn't leave me alone about stealing my gf, sent out homosexual confession letter from his email. Hilarity ensues.

2. DUCKDELICIOUS

I hit my (ex) husband over the head with a 30.06 (?) deer rifle many, many years ago. Then, I grabbed my baby girl and ran. Saw him fall and just left. drove to my moms house the next town over.

Back story. He was a VN vet with PTSD and very violent. He had been in a drug rage for several days. I had been out at the grocery store with the baby and had just come home. I put my little girl down in the den and went to check on him. When I opened the door, he was crouched on the floor pointing the rifle at me. He began ranting that I was having an affair with my doctor (who was about 70 and had been my doctor since I was a kid). He pulled the trigger on the rifle and it made a horrible click…I can still hear it…but it misfired or wasn't loaded. Anyway, I turned to run and he threw it at me and began beating me pretty badly. He got me down and was strangling me. I remember thinking "I'm going to die." Somehow, I got up and tried to run to my little girl…thinking he wouldn't hurt me in front of her. He grabbed a brick (he was a brickmason and had been building a fireplace in our room) and started towards the door saying that "___ would be better off dead than having a whore like me for a mother." I grabbed the rifle and swung it at him and ran.

When I got to my mom's, I was bleeding and battered and didn't even know it. My shirt was completely torn off and I didn't know it. I had a black eye and broken teeth and didn't know it. All I could think of was to get to my mama and make sure my baby was safe.

I had to hide for months after this…I still fear him even though he's now 62 and in a locked facility.

That's my sordid story.

3. CUDAJIM340

I didn't listen to my ex when she needed to talk because I was busy and didn't want to take the time. She hung herself 1/2 hour later.

4. DANIELLEJUICE

My ex boyfriend…He wanted to talk to me and i kept ignoring him. He later shot himself in the head. I still feel like shit.

5. [DELETED]

I had been lazy throughout high school, and never really tried very hard, but just hard enough to maintain a high B/ low A average. Senior year rolls around, and I'm applying to colleges and applying for scholarships and stuff. I speak to our school counselor who tells me that due to my high ACT score, I would be eligible for a $30,000 scholarship, but only if I was in the top 10% of my class. Well, she prints out the rankings and it turns out I'm ranked 15 out of 142 students. Fuck me.

I talk to her about my options regarding my rank, and she says there's really nothing I can do to move up, as it is too late in my high school career. I accept this, and just figure I will have to take out shitloads of loans for school, and end up in debt for years.

So Christmas break comes and goes, and upon returning to school, one of my classmates isn't there, due to a serious mental breakdown right before school let out for the break. She'd always been a little crazy and I figured the breakdown would happen any day. Anyway, I talk to a couple of people about it, and evidently one of my friends, who used to date her, had talked to her and he said that she would be out of school for several weeks to a few months. The conversation continues and we get on the subject of how smart she was and how high her class ranking was, almost to the point of her being valedictorian.

SO I gets me an idea. What if I call her up and take advantage of her state, say a couple of mean things and maybe convince her to switch schools or drop out or something? Once she's gone, I'll be in the top 10% and be eligible for that fat scholarship!

So I go through with the plan. I call her and tell her that I'm calling as a friend, and just wanted to let her know that even though the other students in our class are "grateful she's finally gone" and "glad they don't have to put up with that crazy bitch's shit for a while", I'm still there for her. I say I have to go and hang up.

She kills herself 2 days later.

I got the scholarship.

6. CHIXOR1

In high school, there was a girl in the year below me. She must have had confidence issues, and really wanted to be friends. She was just irritating and clingy, but no matter how mean I was to her, she just wouldn't go away and find other people to annoy. There was a janitors closet with a toilet to the side in the corridor that ran along the assembly hall, so one night after the school musical wrapped up, I told her I had a surprise for her. She let me blindfold her, and lead her to the janitors closet, I had her climb up on top of the toilet seat, and tied her hands to a beam above her head.

Then went home.

7. HACKYSACK

I gave someone a Nickelback CD.

8. HACHALIAH

I made a milk bomb. In my high school cafeteria I took a plastic screw top pint of milk and half emptied it and put some lunch meat in it and screwed the top back on tight. One of my friends found a little heater vent in the cafeteria that we could wiggle off and place down inside without anyone seeing it. We let it sit in there for probably a month during the winter... Mind you the heat was probably constantly on, blowing over the entire bottle.

After a month we could no longer contain ourselves with excitement about what could have grown within that little bottle. So we pulled it out, went over to the freshman section of the cafeteria (we were seniors) and quickly cranked the bottle open and made a dash for the exit into the school's courtyard.

And waited.

No more than thirty seconds later there was a mass exodus of students covering their mouths and trying not to be sick. We wanted to be nowhere near the incident, but from several eyewitness reports, we caused several people to immediately vomit. If nothing else we cleared the cafeteria of about three hundred students in a matter of minutes with the vile odor.

9. [DELETED]

I was dating a girl who had become incredibly dull and boring to me. I decided to dump her, and just before I could get around to doing so, she told me she was pregnant. I was right pissed off, but I held my cool and thought about it, and told her to give me a few days to come to terms with this. I don't want kids, and certainly didn't want to be permanently attached to the vapid girl for life, so after my few days I came back and told her I'd had to think about things, and I'd come to a decision. Now, my GF was a very moralistic person who was totally against abortion as she considered it murder. I told her that I had a family history of severe genetic disorders (lie), and that I'd spoken to my physician and he'd advised against me ever having kids (lie). She initially wanted to 'let God handle it' so she keep to her morals, but I wore her down and convinced her to get an abortion, as that'd be the merciful thing to do. The day came, the deed was done, and she came home. That night, I told her I wanted sex (which she obviously couldn't have, seeing as she'd just had the procedure that day), and guilted her into letting me fuck her in the ass, which was something she swore she'd never do. After that was done, the next morning I told her I couldn't be with her anymore as I felt terrible guilt over the abortion, and she reminded me of that guilt.

10. [DELETED]

I was involved in a drunken fling with a woman I don't find particularly attractive at one point a few years ago. She was one of those women who acts slutty at parties / etc, probably out of a lack of self-esteem and has severe personal issues. For the next two years I proceeded to only be nice to her when I was horny, and then basically ignore her after that, being rude and curt with her whenever we spoke. After two years of this, she got a clue and decided that she ought to move on, quit dealing with me, and got a boyfriend. This annoyed me, since she was my plaything. Keep in mind, I didn't particularly like this woman or find her attractive. So I got in touch with her, and over a period of a couple months went from acting friendly to convincing her I had been a victim of emotional issues during our previous interactions, and that I was better now. I convinced her that I was in love with her. Around this time, she found out that she was pregnant by her boyfriend. I still managed to talk her into leaving him one night when she was over visiting me. She phoned him up while I was sitting on the couch, and broke up with him. I got her to have sex with me that night, and the next day told her I felt awful for making her leave her child's father, that I'd made a mistake, and that we shouldn't see each other again.

He never took her back, and while she at least had a job and some future before all this, now she's a single mother on welfare living in some shit area of town. To this day, she still tries to talk to me sometimes, since she feels that I am one of her 'only real friends'. Her life is what I've made it into.

It makes me smile.

11. CODEEGIRL

I told my biological mother that "just because you got laid doesn't make you a Mom."

12. [DELETED]

When I was like 9 years old, we were making forts in the rainforest and stumbled into this guys pot patch. Being the idiots we were, we took it to our mom and she called the cops. The dude ended up getting 10 years. Now that I understand the context, and how bullshit it all is, I feel perpetually bad for what I did even though it was in ignorance. All the dude wanted to do was grow some fucking plants.

13. HAMSOLO

When I was 19 I began smoking pot, in a bad way, and dropped out. I moved back in with my parents and a buddy moved in to. We spent that winter getting high and chilling. In February he invited a girl over named Sam. I feel in love with her, and he was dating so I figured she was mine. We texted all the time and Kyle, the guy whom moved in, said she really liked me. Sam went back to CO for college and I continued my lust and she continued hers, or so I thought.

An opportunity arose to go to CO to pick up a car and that would put me near her so I was going to visit. Kyle decided to come along for the trip. We made it to Fort Collins, CO and hung out with her. On the second day I told her my feelings and she said she like Kyle. I was like WTF, he's dating another chick. Well it turned out that didn't matter to him and they began to hook up while I was there. I left a day later and he came with.

When we got back to Atlanta, he was forced to move out by my dad for being lazy and I happened upon a job and moved out. We maintained a relationship but I was pissed at him. When my lease ran up he offered to move into a place with me and split the rent. He was living with a different girl and she was there too. I needed revenge for that shit.

I slowly fucked with their relationship, I would bring back in the trash on his days, keep the litter dirty, basically undo all of his chores to piss Kahla off. There was rising tension and they both turned to me for advice, so I gave the worst I could. I told Kyle to be himself because he didn't need her shit and it was better for him. I also told Kahla to be less supportive and with hold sex to try to get him straight.

Well, Kyle left to Bama with his friends one weekend and Kahla went up to her friends too. I found Kyles phone was left at the apartment. I began texting Kahla and saying that he was cheating on her, but from his phone. She got pissed, and her ex showed up thanks to a MySpace message I sent him from her account. They fucked and he managed to get Kyles number. He told Kyle all about how they fucked and he dumped her hard. She tried to say he cheated too and she had messages from him, but he just blew her off.

I had done it, I got her to cheat on him, him to dump her and wreck both of their lives royally. I wasn't done yet however, I had one year of college left and told Kyle I'd help him start anew in anther city if he wanted to move with me to my new job location. He agreed.

I let him live rent free until he found a job. It took him two months but he got a job in retail and started to pay a bit of rent. I begun to work towards getting him fired. I would call the complaint line and leave complaints about how he kept coming on to me while working. He eventually got fired and when he came home and told me, I kicked him out for losing his job. He couldn't find a job and begun to live on the street. Last I heard he was using meth and living in a shelter.

14. BIGTROUBLE77

Rewind to 1992, my freshman year in high school. I'm going to a new private school and wasn't doing too well at making friends. In fact, I was regularly tormented on the bus to school. There was one guy in particular that refused to leave me alone. I'll just cut it short by saying he was extremely antagonistic.

Well, at some point in the school year I got stuck doing a video in history class with this asshole and a few other people. Since we both had the same cameras we were both in charge of video taping the assignment. We had to swap the tape back and forth a few times. At the end of the project he took the tape home and made a final edit to vhs. We submitted the final product and probably got a crappy grade.

This is where it gets interesting… I wanted my original tape back that he had used for editing. After months of harassing him for it he finally gave me the tape. Took it home and decided to show the whole family our video. So it was me, my 5 year old brother and my parents all sitting around the tv… Pop the tape in and quickly realize that this was not the right tape. It opened to someone's basement. We hear someone fidgeting in the background who then enters the frame… nude… jacking off. The fucker video taped himself jacking off.

Once we realized what was happening my mom screamed out in hysterics, hardly able to contain herself. I popped the tape out to spare my little bro, but I immediately realized I had absolute gold in my hand.

The next day I went to school and the usual antagonism started. I very abruptly told him that he had better back off... that I had something he would never want anyone to see... something that was on the tape he gave me. I wouldn't hesitate to give everyone in the school a copy if he didn't back the fuck off.

Well, he said I was full of shit and elevated the antagonism. Long story short, I found a girl that hated him equally and gave her the tape. She made a vhs copy and told our whole grade about it. During lunch break our whole class (about 70+ people) gathered in a class room and played the tape. That day a legend was born: Dancin' Danzis. The entire school found out, all the parents found out and emergency meetings were called to find a way to deal with the situation. Ultimately, the parents and school decided to do nothing, to my great satisfaction. The humiliation I imposed on this loser was so satisfying, I never thought it could feel so good.

I hope he still lives with this shame today.

15. [DELETED]

I made a girl fat... and not by marrying her.

In 2003, my office got a new secretary and a new manager. The secretary, a thin blonde, was a vile she-devil, she wouldn't do anything the staff asked (find info, set up calls, get coffee for anyone who wasn't herself) and the manager wouldn't do anything about it. We joked that her job title was "Internet Quality Control" because she more-or-less sent personal email and played on myspace all day, to which when she overheard, literally went to the manager in tears. Then one day I got a rather large jar of candy as a gift and she just about single-handedly consumed half of it, which pissed me off good. Then later that week, I brought in a dozen donuts and she ate half of them on her own. Upon putting the facts together, our web designer, editor and I decided to fuel the fire in something we called "Operation: Butter-Up", where we each in turn brought in a large bag of candy to fill up my jar as fast as she could empty it. In the course of three months, she put on 20 pounds.

In a years time, she unrecognizable and along with being a crank, she was an idiot and couldn't figure out how. Several times a week we provided her sweets and snacks of every kinds until her boyfriend dumped her, which was about eight months into the project, and she had to get a new wardrobe. We lucked out because she was rather stupid, but we had more diabolical measures lined up if she stopped eating the candy, along the lines of mixing in weight-gainer to the non-dairy creamer she was using and things of that nature. That was 2003 and she's still big to this day, saw her on Facebook and smiled at our handiwork. Still single to this day, and I would like to think that we had something to do with it.

16. I_CREATED_CLIPPY

I created Clippy, the Microsoft Office Assistant.

17. NARCOTRON

In early high school, my best friend and I liked the same girl. This girl was, to put it lightly, "it": drop-dead gorgeous, brilliant, and excelled in all things athletic. Well, my best friend didn't know that I liked her at the same time that he did, and so when the girl and I started to date I didn't tell my best friend for a few months, all the while he was pining after her... I finally told him after a while, and damn was he pissed. I felt like shit. But we didn't stop hanging out even after that—he forgave me pretty quickly, which was always the moment to me that I knew that I could count on this guy like no one else. And which simultaneously made me feel like shit again, because I don't think I would have forgiven him at the time.

Anyway, said girl and I dated for a long time, and it turned out to be the most intense, emotionally taxing, fucked-up relationship I've ever had. Whenever my friend and I joke about it now, we always comment on how I ended up saving him in the end. Karma's a bitch.

18. BILLY357

One night I was at a friend's house party and living it up as usual. My friend (owner of said house) introduces all of us to a real asshole friend of his. This guy was bragging how he just stole this teens purse over at the local burger joint. I'm not usually one to get involved or be of high morals but I've got a thing against stealing directly from innocent people (stores and stuff were ok at the time). So later on that night when this guy is tipsy I get him in a part of this house, just me and him.

I've been brooding all night over this guy so I'm really pissed (and drunk) by this time. When he turns around I'm sticking my Smith & Wesson 9mm pistol in his face. He literally shits himself but I don't care. I take this girl's purse from him and then make him give me his wallet ($353.00 in cash, a license, some credit cards, not much else) and all the valuable shit he has on him (gold necklace, class ring, iPod, and some weird demonic ring). I tell him to never grace the presence of me or my house friend again as I cock the hammer back on the gun, at this he pisses himself. I make him leave the party right then and there (I left him his car keys). Later on when I sobered up I drive over to the local big boxmart and get a box out back of the dumpster. I note this girl's license address and label the box as such. I then put in her purse with all her belongings plus this guys valuables (minus the demon ring) and his $353.00. I then proceed to tape it up and ship it out to her. I then went downtown and gave this guy's wallet to a homeless man and told him to live it up on his credit cards.

19. ALLOTRIOPHAGY

At primary school in Scotland, when I was about 9 years old, a new boy, Peter, joined the school. His family had moved up from England. He was a bit odd but we go on well enough.

After a few months, he stayed over at my house as part of a sleep-over. A few days later I was looking for something in my toy cupboard and found a pair of Y-front underpants. They weren't mine. They were Peter's and they had the most enormous streak of shite in them.

I gingerly put them into a clear plastic folder and the next day at school during break time I charged people 5 pence to look at them. Then someone took them from me and everybody chased Peter around the play ground, with his dirty knickers waving in the air.

20. MAGICFROG9

Once I filled some empty corona bottles with urine, recapped them and left near the recycling bin across the street. I was hoping to see some vagrant drink one, there's bums drinking beer outside my apartment at all hours. After an hour, someone did grab those bottles, but instead of chugging one down right away, they just put them in a backpack and scurried off. Dude didn't look much like a hobo, very unsatisfactory experience overall. Afterwards I felt guilt.

21. YEAHIMAWFUL69

This honestly still makes me LOATHE MYSELF, even years later. Okay this will be short and (not so) sweet: I let my boyfriend who I was living with at the time eat me out, right after I had gotten home from cheating on him. Worst part… the guy had cum inside me. Despicable me.

22. KARMANAUT

A girl was very obviously cheating off of me on an exam so I filled out my test with all of the wrong answers and made it pretty clear so that she could see them.

After she turned in her test while I "checked my answers", I erased the wrong ones and put in my own answers. She was quite surprised to recieve a 0 on the final, whereas I got an A. She failed the class.

23. CULERO

I took a girl's midterm (it was a large class) in exchange for a BJ. I got one for taking it, and another after she got an A on it.

24. TANMANDU

Me and some friend took fishing line to school one day in high school. We got into groups of two. One person held the base and the other took the loose end. We stood in the hall and "weaved" in and out of people. Thus creating a massive "web" of sorts. Mass confused followed and it was pretty damn cool.

As fun as it sounds, we ended up cutting a lot of people with it and caused a older teacher to fall and break a wrist. i know of lots of students that still have scars from it.

Somehow, it never got back to us and they are still wondering who done it. It bothers me to this day.

P.S. it still was pretty damn cool.

25. DAWN_OF_THE_DEAF

Back in college, my roommate was gross, she used to not take a shower for over a week, she would wear the same pair of trousers for a whole month, she would leave all her dirty (DIRRRTY) clothes on top of my freshly washed towels, she used to smoke in our tiny bathroom, and a long gross etcetera.

She used hair-removing cream to get rid of the hair in her armpits (which is totally fine) but afterwards she wouldn't clean the sink or remove the cream bits filled up with armpit hair. Day after day, the same cream bits were there, I used to have nightmares when I thought of how gross those were. Every time I needed to brush my teeth, I felt like throwing up. After warning her kindly twice, and seeing however that she still wouldn't do anything about it, one day I decided I would, from that moment on, clean it myself. Am I not a dream roommate? I used to leave the sink spotless, her toothbrush did the trick very well.

I used again this cleaning method when my brother kept sprinkling the toilet bowl instead of leaving it up while taking a piss. He has damn good teeth, though.

26. REAL_DR_EVIL

Last year, I started my 4th year of med school and got transferred to a public hospital to finish my studies, near the surrounding areas of the hospital there is a slum where thieves hide. One morning going to class I got mugged at gunpoint by two guys, they took my backpack, phone, wallet and shoes, in my backpack I had a stethoscope that was quite expensive and that my dad gave to me as a gift when I started to go to the hospital, they took it just because they wanted to screw me as I told them that I only had that and a notepad in there, I was really pissed of at the time but I could do nothing if I wanted to live.

After some months I started to do assist on the ER and one night one of the thieves came in with a laceration on his cheeks, now I've got to say that I live in South America and it's a common rule that at night medics go to sleep and when a patient appears on the ER we have to evaluate them and then call the medic, which isn't legal at all. As I was alone in the ER at that time, nurses were on their station, and I wanted to do some harm to this guy I put my act as a medic and started to suture him, which is something that I know how to do but it'll leave a scar as I'm not very good at it yet, but leaving a scar wasn't good enough for that scum, I wanted him to suffer so first of all I diluted the lidocaine with saline solution and applied it in a way that I was going to hurt him so I did my best to suture him without cleaning the wound, using not sterile gloves and touching all the suture materials with my bare hands before using them, when I was done with the suture I didn't even register him on the books and send him off.

One week later I heard that the same guy came back with a nasty infection that required antibiotics and should have left him with a nice crater on his face but as he didn't come back the same day that I'm doing my practices I don't know what happened to him.

27. KNIFE_NINJA

The day after "Harry Potter and the Half Blood Prince" came out I played counter strike all day with the handle SNAPE KILLS DUMBLEDORE.

28. MICKOLASJAE

When I was in preschool, I apparently put glue on the entire class's seats. Didn't get to go to recess, stayed in and broke crayons in half and threw them at the flagpole.

29. MURFTRIXON

When I was 11, I gave a kid (he was 9) a pretzel stick I had stuck in my cat's litter box.

30. FRICKTHEBREH

This rather large kid had a habit of bullying me out of my Oreo cookies at lunch everyday. When you're in second grade, this begins to take a toll on you. Finally, I went to my dad about it and he concocted quite the plan, halfway inspired by something that happened in a children's book he read to me at the time. He took one of the cookies, hollowed out the center of the cream, and put a ton of Tabasco sauce in there. When he resealed it, he left cream on the outer edges so it still looked normal. Needless to say, I was incredibly excited about what was going to happen the next day at lunch.

The next day, he actually wasn't bullying me out of the cookies…he must've had a change of heart for once. But A.) I wasn't going to eat that shit and B.) I still wanted to discourage him from ever stealing them again. So without provocation, I took the bag of Oreos and offered them up to him for free, on my own behalf. Delighted, he took a bite of the atomic one and immediately started coughing to the point where he was pseudo-choking. He ran to the bathroom, vomited (I believe), and I didn't see him for the rest of the day.

He never stole my cookies again…I still smile about it to this day.

31. [DELETED]

I use Craigslist to post very friendly and sensitive sounding advertisements to young women who are in a financial bind. When they get in touch with me I gently talk to them over a series of emails until I get to the point where I suggest sex for money. I ignore all of the ones who are enthusiastic, even the first-timers, and only go for the girls who are repulsed by the idea but are in dire financial straits. I tell them that it's not really prostitution, that it's just a boyfriend who helps you out, etc. etc.

Then, when I penetrate them, I tell them that I've just turned them into a whore for the rest of their lives. I love watching the emotional anguish. I do this in a way that seems innocent, like I offhand say "Oh man I've never actually turned a girl into a prostitute before" in a really excited tone of voice — this avoids them hating me, which is good because it avoids retribution, and it makes them turn their loathing in on themselves instead of having a perpetrator to blame.

I get off on the huge power trip of psychologically fucking with these innocent girls.

After they leave in emotional shambles, I friend the girls under a fake Facebook account (9 times out of 10 I can discover the girl's real name).

My favorite score was a girl who used to post 5-10 times a day on Facebook and immediately after seeing me only posts once or twice a month. She started crying so I told her she could stop but she wouldn't be paid — I did this so that she would allow me to continue fucking her while she cried which is one of my all-time best sexual experiences.

In effect, I do the S of S&M for real, not in a controlled consensual environment. I've done this to about 40-50 girls mostly aged 18-24. I also use psychological tactics to reduce the price as much as possible, and usually spend around $100-150 to do this to each girl. I also videotape the sessions with a hidden camera and the camera has never been caught.

I am an extremely successful person career-wise, have a lot of great friends, great relationships (I don't do this when I'm dating), and most people consider me to be an excellent upstanding member of society. But I sometimes wonder if this psychopathic side of me will end up destroying my life someday.

32. [DELETED]

I told my step brother i would give him all my nes games if he took a handful of hard packed dirt that was mixed with chainsaw oil, gas, dead deer stuff, dead fish stuff, from the back of my dads truck, and chew it slowly while i counted to 10. he did it… I laughed and gave him nothing.

33. ALLI3THEENIGMA

In middle school, there was this girl that tortured the living shit out of me and got all of my "friends" to turn against me by going into the school after hours for some kind of school event, jimmying my locker open and throwing the contents around the hallway.

Well, I was so traumatized by those girls that I decided to go to High School in a different town and my senior year I started dating a guy that graduated a year earlier from a private school in the area. In a bizarre twist of fate, I learned that this girl now attended that same school and, since the classes were so small, they had large "cubbies" in the Senior lounge with their names marked on them, rather than lockers. My boyfriend and I were helping his Dad one night at the school (he's the head of the English dept.) and I snuck into the senior lounge and got my revenge by breaking her reading glasses, stealing a single shoe left in there, and throwing around some of her shit. Right afterwards I felt bad and confessed what I did to my boyfriend but he thought I deserved a little revenge.

34. PYCHOMEDIC

I put dog shit in a slightly 'special' kid's sandwich. He then took a bite out of the sandwich and choked on it.

(It was the dry & hard kind of dog shit. No he didn't die or anything.)

35. [DELETED]

I once told a girl that she was going to end up alone just like her mother, who was essentially her best friend and had gone through two divorces. I was pissed at the time and I really regret saying it, but it was completely true.

36. SUNSHINE-X

We shot our best friend in the chest.

We were kids, and a friends dad was a cop. Let's call him TS. He had rifles and hand guns in the house, and he didn't lock them up. We used to shoot them in the basement at a log pile and enjoy the crazy ricochets.

Anyhow, one day, one of my friends played a mean practical joke on me. We'd just returned from the hospital (visiting a sick classmate, broken leg) and the nurses had given us empty syringes to use as water pistols to fight with him. It was a blast. We got back to my friend TS's place, where we would be sleeping over. We kept playing with the syringe water pistols and one of the guys, let's call him JM, comes down to the basement with a syringe full of something white. He shot it on my face and hair, and yelled "that was my cum!!". I wasn't impressed, and ran to wash it off. When we came back, we decided that what he'd done deserved some form of punishment and he agreed to participate in a plan we'd been working on.

We wanted to shoot someone, while wearing a bulletproof vest. No one wanted to be shot of course, but he agreed that his dick-move would be forgiven if he'd be the "victim".

We dressed him up in the old vest of TS's dad, placed a skidoo helmet backwards on his head, tied his hands behind him (so he wouldn't accidentally take it in the arm), and stood him at the end of the basement.

We set up a pile of logs to use to steady the rifle, took aim, and fired a .22 long into his chest.

He didn't like it at all, and cried/whined a lot. We took the vest off, and other than some (later) bruising, he was fine.

Anyhow, looking back, it could have ended very poorly. I should never have done this.

37. SANDKILLER

In 2008, I went to S. Korea for a few months for work. Most of the time, I was not in Seoul, but I was scheduled to spend my last week there.

On the night before I left for Seoul, I stayed up late and IMed people back home, including my best friend since the first grade (let's call him Brian). I was his best man when he got married in 2006, and he was supposed to be mine when the time came. Little foreshadowing there, I guess.

He kept saying weird stuff. Not weird stuff, exactly. He just wasn't himself. He also kept encouraging me to go out and have some fun instead of sitting in my room and IMing. I kept telling him that I had a lot of stuff to do the next day before leaving, that the night life in the city where I was staying wasn't anything to write home about, that I didn't speak the language, etc, but he just wouldn't drop it. Finally I got a little angry at him about it and told him he was being rude. He was profoundly and unusually apologetic and very intent on making sure that we were still pals. I assured him that we were and apologized for losing my temper. Then I went to bed.

About a day later, in Seoul, I went to an internet cafe (PC bang) with one of the 15 or so other people I was traveling with. In my inbox was an email my mother had sent me, telling me to call home, she had news about Brian. I thought, "huh, was he in an accident or something?" All the computers in those places have headsets and skype, so I logged into my account and called.

When my mother answered, she told me that he'd been accidentally shot and killed in the gun store where he worked. I don't remember what I said exactly at that point. She asked me if I wanted to speak to my father, and I said no. After I hung up, I put my head on the desk and wept.

After I partially regained composure, I called the travel agency to see if there was any way for me to come home a few days early. There wasn't. Everything was booked solid. I emailed all of our old friends and gave them the news. Then I and the guy I was with went back to the hotel room for a nap. I'm sure he noticed that I'd been crying, but he was the kind of guy who minded his own business, and he never asked what was wrong.

I didn't tell anyone. I didn't know any of the people I was with all that well, and there was nothing they could do about it, anyway. I just buried it and went through the motions of what I needed to do for the remainder of the trip.

That night, when I checked my email again, I found an article that one of my friends had googled up. According to it, the local PD in Brian's town had ruled his death a suicide. On his lunch break, he'd gone out to his car and shot himself.

I found out after I got home that his parents, brother, wife, and in-laws refused to believe it. There was no note, for one thing. But I knew it was true. That weird IM conversation we'd had convinced me of it more than anything. But I also just knew him. He never acted depressed or despondent, but there had been hints that he was in trouble for months. They just didn't stand out until after the fact.

I of course missed the funeral. I was probably singing karaoke or something. I never told Brian's family about our final conversation or that I think it was probably a suicide. In fact, I've scarcely spoken to them at all. My mother called me once and told me that Brian's wife had relayed a message to her through his parents that she was "willing" to talk about it if I wanted to. I never called her. When I was visiting my parents' for Christmas a year ago I ran into his dad at the grocery store. He told me they'd like to have me over to the house for dinner sometime. I was polite but noncommittal, and it never happened.

It wasn't for some months that it occurred to me

that if I ever got married, I'd have to find a new best man. When I thought of it, I cried.

I'm not sure where I'm going with this. None of it is particularly evil. It's a little dickish and cowardly that I never tried to comfort his widow or family or even acknowledged to anyone that his death upset me. I don't think his death is my fault. I don't feel guilty about it in that way. I think I just saw a couple of other stories about suicide and decided to unburden myself a little.

38. INFECTUS

One night long ago, my band was playing a gig in a local bar. Before the show I saw a beautiful girl standing behind a chest-height counter. Every time I passed by to get a drink or use the can, she was watching me with a big smile that really lit up her face. I decided that after the gig, I was going to chat her up and hopefully get her number.

Well, we played our set and then started packing up our gear. Somehow in the shuffle I lost sight of her and we ended up leaving. As we were getting in our vehicles, I heard a voice call out from the parking lot: "Hey, you know you're cute?" she called out. My buddies and I turned around to see the girl that had been smiling at me earlier. She was standing with a few friends and I could now see that she was a plus sized girl. Without skipping a beat, I called out "Hey, you know you're fat?" and we left.

You can't make me feel any lower about it than I already felt as soon as those words left my lips. God, what an asshole. That was almost 20 years ago and it still haunts me.

39. ELVISLIVESON

Not sure this is evil as more of an accident of confusion, once many years ago i basically took a huge shit on my buddy's face while he was passed out. we had been drinking for hours and didn't make it home that night so we crashed in a dark alley on the way. he woke up thinking he had somehow landed on someone else's shit. i secretly laughed all day. guess that was evil.

40. RIKATEE

When living at home, me and my bro would squabble a lot, just little stuff.

However, for some reason I cannot remember, I decided to knock it up a notch BAM! by sliding chicken into the tubing of his blinds on the window next to his bed.

After a few days the smells started and after a few more the flies came and after a few days I hear this almighty scream, the likes no one can generate unless they wake up to maggots falling from their window onto their sleeping face – which, funnily enough, was what happened to him.

41. NICKSAUCE

New year's eve 2002, if I recall correctly, just after midnight had passed. People were announcing "First X of

2003" for everything they did. I don't know why, but I decided to go for "First person to kick Joe in the balls in 2003!". He was very close to hooking up with a cute girl that night, but instead he spent the rest of the night lying down in pain.

42. BXCELLENT

When I was a student, our neighbors were complete assholes (also students, but from a poly, not the uni) and had enjoyed pelting our house with eggs, dumping garbage in our back-yard, etc.

One night, after much drinking, a friend and I found a gas pipe on the way home that we were sure was destined for better things, so we took it home. As we arrive home, we went around the back of our house and discovered a number of garbage bags, and a bag of quick-drying cement. So we dumped the garbage bags into our neighbors yard against their back door, and covered it with the cement.

Then we pushed the gas pipe through our bathroom window, connected a hose to the bath tap, and dosed the cement with water through the gas pipe. We then attached the gas pipe to the top of one of our friends cars and went to sleep.

The next morning, the police were with the neighbors and they had no idea what had happened. Their back yard was solid cement and garbage and they couldn't open the back door. Their landlord arrived and threw them out and the police never even asked us one question. Revenge was sweet, but sometimes, when I think of it, I do feel the slightest pang of guilt.

We shouldn't have put the gas pipe on our friends car, cos he got fined for it....

43. SCOTTWUZHEAR

In 6th grade I was friends with a guy named Jason. Jason had begun obsessing over this girl in our class, so I thought I'd play a prank on him. Before Thanksgiving break started, I found out the girl's email address and AIM screen name, and then I pretty much guessed her secret question answers on hotmail, thus I had access to her email and AIM screen name.

All throughout Thanksgiving break, I pretended to be the girl on AIM, talking to Jason every day. The funniest thing was, he would often call me immediately after talking to her on AIM for hours and excitedly tell me all the things that they talked about.

As it got closer to time to go back to school, I started pressing towards forming a relationship between the girl and Jason (and he relayed this to me via phone). Eventually, 'the girl' told him "If you really like me, I want you to just come up to me on Monday and smack my ass in front of everyone. I think it will be hot, and I'll know you're really into me!" Needless to say, Jason was nervous but excited. He kept asking me if he should really do it or not, and I would always tell him yes or else he risks losing the girl of his dreams.

So, Monday rolls around and I've told almost all the guys in class (I didn't tell the girls because I figured they'd tell the girl whose ass he was supposed to smack.) We're all watching him like a hawk, waiting for him to do it. The bell is due to ring in less than two minutes and the girl is at her locker. Jason slowly goes up behind her, and I see him cock his hand back.

Then I chickened out. I realized that were the girl to find out that I wanted a guy to smack her ass, I might get in some stupid sexual harassment trouble. I grabbed his hand and stopped him and said "Jason.. it was all fake."

"Huh?" "Yeah, it was all fake. I hacked her AIM screen name and posed as her for 2 weeks."

And then I saw in his eyes a look of sheer horror. His world was crashing down, and it couldn't possibly be true. I felt horrible, but I also felt like laughing. The girl eventually found out and absolutely hated me for it, and Jason hated me as well, although we still talked and ate lunch together.

Fast Forward: Jason eventually gets over her after a long period of being obsessed (he'd even whistle out loud every time she walked into the classroom). The girl stopped hating me and thought that the whole ordeal was hilarious. In high school, Jason pretty much forgives me and chuckles over the ordeal, but changes his mind after another girl shoots him down and tells him he looks like a 'concentration camp jew'.

I still feel evil, but it was executed so perfectly that I can't help but be amazed and chuckle occasionally.

44. TFJ

I created the font known as Comic Sans.

45. DR_LANDREW_5

There was once this kid who used to give me a hard time on the bus when I was in High School. even as seniors he never let up on me for some reason. He was in my english class and was barely passing with D's. One day he squirted an entire bottle of ketchup in my locker and all over my new Nike jacket my mom had gotten me. We didn't have money for brand name things so this jacket was big deal to me. He completely ruined it. Well towards the end of our senior year, I volunteered to help out as a sound tech for the school play.

Our English teacher was also the drama teacher and she would let the sound people use her classroom to keep our stuff in. I found myself alone in her class one night during practice and noticed her grade book opened with our english class grades in front of me. She actually wrote everything in her grade book in pencil. So I quickly grabbed a pencil and changed some of his grades. A month later, he failed that class (barely) and didn't graduate with our class. He had to go back the next year and graduate a year later. It was quite the shock of our little town b/c he was a very popular boy from a very popular family. I still don't feel bad about it but I realize it was evil.

46. LEEROYJ

My old roommate used to get mad and hit me when things didn't go his way or got drunk, so I did the unforgivable… He was a black guy and had bottles of Lubriderm lotion laying around for his ashy skin. He used to rub it all over his body and face. I kept thinking of ways to get revenge and it randomly came to me. I masturbated into a cup and poured it into the bottle of lotion… A day later he comes out of his room with his shirt off, acting like a douche bag and begins rubbing the lotion on his chest and face… I couldn't contain myself. My other roommate told me, that he in fact, did the same thing after I told him of my fiasco.

47. L0STM4N

There was a kid who lived in the house whos back yard butted up against my grandparents. This kid was the type who would play with you no matter wtf evil shit you did to him. I guess it was my own ego looking for an outlet since I was pretty picked on growing up. This kid wanted to play with me and my cousin. We let him come over to my grandparents back yard. We played cowboys and indians. Me and my cousin tied the kid to a pole near the swing set, pantsed him, and went inside to watch TV. I don't remember how long it was but his father came home only to see his son tied to a pole with his pants at his ankles. I got my ass whipped for it by my grandfather. I still feel bad about it to this day.

48. JONSEY737

In grade 7 I knew a guys gym locker combination, I took a sample maxi pad that came in the mail and put it in there. While he was getting changed for gym class it fell on the floor. Other students proceeded to kick it out of the locker room and around the gymnasium floor, all the while making fun of him for being a girl. He didn't come back the next day and never returned. I'm not sure if it was related or not but I feel a bit guilty.

49. NOTGOODNOTGOODATALL

Blamed kid down the street for breaking a window I broke at my house. Kid got a beating for it. So did I when it was discovered that I lied. (~6yo)

Flattened every tire on some guys raised pick-up (with giant tires) by unscrewing the pin in the valve. Woke up the next morning to see his car on blocks — He had taken the tires to the shop. (~11yo)

Shot kids (when I was 11yo) with blow darts made from straws and cactus thorns. We also told them the tips had been dipped in black widow poison.

Accidentally broke someones arm after I intentionally pushed him into a wall while we were chasing a ball in gym class. (13yo).

Threw full 7-11 cup of Dr. Pepper over the top of stall onto some unsuspecting guy taking a shit. (~16yo)

Filled a bucket of every rotten old food and drink in my college fridge (plus several bottles of cologne and about 30 slugs), let it sit in the sun for about 2 weeks until it became a loaf, rehydrated it with rotten milk and beer, filled plastic bags full of the material, then drove around looking for people to hit with the dank. At least 4 people had direct hits scored on them. This stuff was awful smelling. Really awful. I still feel tremendous guilt over this one. (~19yo) Note: With the exception of the first one, the rest were the result of bad influences. That doesn't excuse it, I know.

50. SCHAFER09

In third grade, I cheated on my history exam. In fourth grade, I stole my uncle Max's toupee and I glued it on my face when I was Moses in my Hebrew School play. In fifth grade, I knocked my sister Edie down the stairs and I blamed it on the dog...When my mom sent me to the summer camp for fat kids and then they served lunch I got nuts and I pigged out and they kicked me out...But the worst thing I ever done — I mixed a pot of fake puke at home and then I went to this movie theater, hid the puke in my jacket, climbed up to the balcony and then, t-t-then, I made a noise like this: hua-hua-hua-huaaaaaaa — and then I dumped it over the side, all over the people in the audience. And then, this was horrible, all the people started getting sick and throwing up all over each other. I never felt so bad in my entire life.

51. PROPMONKEY

Live in a small vacation town that's quite abandoned in the off season. One particular morning, while waiting for the bus, I decided that I would skip a fine opportunity to attend school. Since my Dad started work in over an hour, I couldn't go home after skipping, so instead I went up the block and strolled about town. Once I had traveled a couple blocks of eerily silent shore town, I found the nicest outdoor shower I came across, one with a bench, and hung out inside it smoking a few joints (back then I stocked my wallet like a portable smoking kit).

Now, before I go to the bus stop, I'll usually make sure to drop a load, since my bus ride was around 45 minutes long. This particular morning I neglected to do so, and paid the price accordingly. Waiting for my father to leave for work had me pushing well beyond my shit-withholding threshold, and I was cold and high. The solution came easily enough in the form of an unlocked door behind what I assume to be a family's winter home and summer-rental. I'm sure it was seasonal as the house itself was abandoned, and the unlocked door led me into a small, 4×6 ft laundry room, something of a rarity. Doing the logical thing I searched my backpack for tissues, found some, and dropped my pants thusly, before jumping onto the washing machine and dispensing into it a really nice shit. When I say nice shit, it comes from the heart. You know those rare, amazing shits where it slides right out, intact and large? And then the feeling of success after having wiped your ass and finding that the toilet paper, or in this case tissue, has almost nothing on it? So I shit in the random house's washing machine, and nearly thirty seconds later was gagging at the smell that accumulated in such a small space, something I never anticipated. Being that this was a seasonal residence, I consider the occupants lucky, as, for most people the shit would have lingered, slowly building it's lethality, for months.

I was pleasantly surprised a week after my deed.

While walking to the beach to smoke a blunt, I saw a bald middle aged man with a scowl hosing out the room I was so recently defecating in. Interesting to note was the large amount of children's toys in their yard. I imagine the family happily pulling up in the driveway, ready to enjoy a weekend away from it all, and then the expression on the face of the little girl who discovered my joyful shit. I like to think she looks like Dora the Explorer in real life.

I feel the most evil part of this is that it would've actually been a shorter trip were I to shit on the beach and bury it.

52. [DELETED]

Introduced Yeast Infections to both of my female house mates as punishment for not paying rent on time…

53. LORANIS

One of my friends in High School had cancer, it was the really bad kind and he had been fighting it pretty much his entire life most of the time the doctors kept telling his mom that he would not make it.

I treated him like shit, everyone else was incredibly nice to him because of the cancer but i always acted like he was an annoying little bother. The funny thing is, I think he kind of appreciated it, I was the only person in his life outside of his family who did not pity him.

He was so pathetic, even when the doctors told him

he had finally beat the cancer he had no clue what to do with his life, he spent so long assuming he would die before turning 18 that he had absolutely no ambition other than playing games on the computer he got from the make a wish foundation.

He died a few months after turning 18, I was studying abroad but my parents told me he tried calling me a week before he died even though he knew i was not there.

I feel like the lowest form of scum on earth for the way i treated him, still have not been able to bring myself to visit his grave to apologize, maybe some day.

I am truly sorry Dave, i should of been a better friend to you.

54. JOOES

I bullied someone who later killed themselves. I've never known the reason why she killed herself, but there is always that thought in the back of my head that makes me think I was responsible.

55. NOT_ME10

I beat the shit out of my 7-year-old-son and told him I wished he'd never been born. I beat him pretty regularly until he got 10 or so, and I got a cold, hard look at myself.

It was the most despicable thing I ever did in my life, and I have regretted it every day. He turned out pretty psychopathic, by the way, but probably not as psychopathic as I was at the time. I'm still pretty nuts. If I was completely sane I'd probably kill myself.

56. PHIL_COLLINS

My friend and I went swimming, there was a pretty good crowd at the pond in our neighborhood, but not the most crowded I have ever seen it. Anyway I was jumping off this little rock outcropping on the far side of the pond when I saw my buddy cramp up and go under.

I felt totally helpless when I heard his gargled scream for help as he went under and he didn't come back up. My friend died right in front of me. I saw it with my own two eyes. I jumped in and tried to get there as quickly as possible. I just couldn't make it and I felt really bad for years about the whole thing. I think I really channeled my anger to the other people there that day, especially one jock dickhead kid who was a little closer than me. He might have been able to make a difference, but I think I just blamed him as a way of coping.

Years later I still blamed this guy for my friends death, and I invited him to a concert (front row seats) had the stagehands spotlight him and sang In the air tonight. You should have seen his face!

57. MILDFIRE

This might not be that evil, but I feel bad about it now. IP Relay is, or was, an operator service that helps blind, deaf, and/or mute people communicate. For example, let's say a mute person wants to order a pizza. They open the IP Relay chat window, type the number they wish to connect to, and the operator will dial the number, say whatever the person types, and type back what the pizza place says.

Well, during my freshman year of college we would get drunk in the dorms and type the most God-awful things in the chat window, and have the operator call another one of us who was in the room. The operator would have to say whatever we typed, and you would not believe the things they would say without laughing or getting angry and disconnecting us. Hilarious at the time, but I regret it now.

58. GUANWHO

I was working as a security guard at a grocery store when I noticed someone edging towards the exit with a basket of groceries. I made myself scarce but kept an eye on the guy and sure enough he bolted out the exit. I ran him down and tackled him in the parking lot. the police were on their way when I noticed that the guy was trying to steal baby formula. I don't know who the real asshole is in this situation but I've always felt wretched about it. Thankfully I don't work security anymore.

59. THROWAWAYACCOUNT910

I'm an asshole to everyone (even my friends). Overall I'm a pretty miserable person and I try to drag people down to my level by being a jerk at all times.

I do this on purpose to hide my insecurities and overall unhappiness. I've made people feel bad about themselves and for that I am greatly sorry. Regardless of the fact that I know this, I do not change.

60. THROWAWAY30225811

I have three very evil things. This is my throwaway account. Here's my list:

In fourth grade, we were supposed to do a page from this workbook every day, and on the last week of school, we were supposed to hand in the whole book to the teacher. I didn't do any of it, but another kid apparently forgot to put his name on his, so the teacher assumed it was mine and not his. I took the grade and never said anything.

I got in a car accident in college when some old lady rear ended me and totaled my car. At the scene of the accident, the old lady started swearing at me, telling me that the whole accident was my fault. It clearly was not my fault as I was at a stop sign, waiting to turn left when she plowed into me full speed from behind. The insurance company sided with me and agreed that the woman was 100% at fault. I had a minor neck injury from the accident, but it wasn't a huge deal, and I had insurance, so my treatment was covered, but since she was so evil and mean to me at the scene of the accident, I decided to sue her. I got $14,000, but I feel bad about it to this day.

I had a boyfriend in high school who was that classic way too nice guy. I cheated on him twice (once right in front of him), but he took me back anyways.

A few years into our relationship, I convinced him to move two hours away with me because I got into a college in another state and I didn't want to go alone. He applied to a nearby college and got in. We got a lease to a fairly expensive apartment which we planned to move into together. A few weeks before we were set to move, I met another guy, we hooked up a few times, and then I got an acceptance letter from another (much better) college near where I was living at the time. I made the choice not to go with my boyfriend, leaving him with an expensive lease that he couldn't afford on his own. I then broke up with him, and started dating the other guy that I hooked up with. I heard from friends that my ex ended up having a hard time making friends in the new state, couldn't afford his rent, went way into debt, and was finally forced to move back in with his parents.

I feel really bad for everything that I've done, and I would really like to make it up somehow, but I really have no idea how to do that. I just try to be a good person now, but I really regret the things I did in the past.

61. INDUBITABLENESS

I beat the ever loving shit out of my brother one time. It was the last fight we ever got in. He was bleeding out of his ears and mouth and nose.

I felt so bad I ended up crying about it for almost three days. The fucker never pulled a knife on me again though.

62. LOGGINGOFF

I had a couple hours between classes in college one semester, so I would usually go down to the computer lab and dick around on the internet until I had to go. The browsers there are set up to remove all your history when you close it, but at the comp I sat down at the person before me seemed to have frozen up firefox or something because when I clicked it it asked me if I wanted to reload everything. Out of habit I clicked yes, and this guy's facebook pops up, still logged in and everything.

At first, I just posted a message on his facebook status that told him he should be more careful and log off his stuff when on public computers and clicked off to some other site.

A little while later, I went to check my email, and this same guy from facebook was logged in to his.

I was able to not snoop the first time I was left with an open account, but not a second time. I started reading through some select emails with interesting subjects, and find some between him and a ton of guys from craigslist, sharing dick pics and setting up times to meet NSA.

Basically, from reading his emails, and clicking back to his facebook, I get the idea that this guy has a gf, listed as straight, and is cheating on her with a bunch of random dudes.

This pisses me the fuck off, so I find the worst, most graphic emails and forward them to his entire contact list, which included the folders friends, family, professors and work. I even took the time to resend them all individually when some of them failed due to some people's addresses being invalid.

The entire time I did this my heart was pounding.

Looking back, there was a chance the girlfriend may have known about the craigslist stuff, and was ok with it, but I dunno... just from the way they were written, and the front page of his fb, it just didn't seem that way.

A while after that I was so scared karma would come around and kick my ass, but nothing ever happened...I was expecting this huge investigation from the school or something, but... nothing. I guess I've written it off as a moment where I myself was acting as this dude's karma, but I still feel like shit about it. I kind of wonder still how much I must have fucked up this dude's life.

63. GEORGE_LUCAS

Where do I begin?

64. THINGSNEVERSAID

Once, I decided, against my better judgement, to date one of my closest friends. It actually worked out rather well, and we ended up going out for about four years. Eventually, things began to fracture, and serious irreconcilable problems began to arise that forced us apart. She was absolutely devastated; it completely tore her apart.

Within a month, I had sex with her best friend.

65. ARCTURUS_MENGSK

I left Sarah Kerrigan to die on Tarsonis to lure the Protoss into attacking the Zerg.

Jim, I'm sorry.

66. [DELETED]

There was a girl in college, we will call her jessica. She flirted with me a lot but I had a GF, and I was not interested in her at all, she always talked about the men she hardly knew coming over to fuck in her in house. Either way, one day I broke up with my girl and showed up to class quite happy, Jessica asked why I was so happy and I told her. She asked if I wanted to go on a date. I told her something about just broke up blah blah not looking to get back into dating blah blah. She then stepped closer and said, "Well maybe we can just get together to fuck?" I then gave her a "too soon" kind of answer.

Fast forward a few weeks, we were in this DB class or some shit, either way, I never paid attention and had no idea what the fuck was going on. For the final we had to put some db together and take screenshots of it and then turn it in. I gave it a try but couldn't get it to work. My friends in the class were just as lost as I was. Jessica knew what she was doing. and had worked out a deal with most of the guys in class — they give her 30 bucks she hands the stuff over to them.

I looked at one friend and said, "I can take care of this."

So I slid my chair over to her and placed my hand on her leg between her knee and hip, (did I mention she was a little big and ugly?) and said, "Hey, how you doin?" she replied, "Fine, just finishing this up."

I then start in, "Yea this stuff sucks. So, that offer to come out and um.. you know," as I start rubbing her leg, "come out for some uh.. fun *wink* — does that still stand?"

She says, "Yea, why?"

"Well," I said, "I would love to come out tonight, but I'm going to be stuck here all day trying to do this shit, and by the time I'm done, I won't have the energy for anything else," and while I'm saying these things, my hand is working slowly and softly closer to her inner thigh. She then chimes in with, "Well, you could have a copy of mine, and then you could make it out faster."

"That would be great," I say. "You almost done?" Jessica then says, "Yea, just finished wheres your thumb drive?" I handed over my thumb drive, she copied her final onto it and handed it over. I then stood up and proclaimed to my friends that we're going to pay her for it. "Hey guys, I've got it!" I said. She looked at me confused and said, "Wait, you're still coming over tonight night, right?"

I turned to her and said, "Of course I am." I then made the changes to make it look like my work and passed my thumb drive around to all my friends so they could do the same. We turned in "our work" and started heading for the door.

As I started to leave, she asked again. "You're still coming out tonight right?" I told her, "Of course, I'm just going to grab lunch with the guys, I will call you about 6 so you can tell me how to get to your place."

I walked out the door, went to lunch with the guys. After lunch, me and a friend went back to his place and place and played video games all day and I then went home. Jessica never hit on me again, and didn't talk to me for months after that.

7

44 Employers Reveal The Most Ludicrous Things They've Ever Read On A Resumé

Having a hard time finding a job? Well rest assured, someone out there is having a harder time than you. From r/AskReddit.

---

1. PI_BEER

    "None of my references really like me, so please don't believe what they say."

2. WRAITH_MAJESTIC

    "I am in the top 2% of programmers." No explanation of how that is determined… I should have asked. ?

3. MIKEBSNYDER

"Grate communication and attention to details" on a resume I looked at maybe an hour ago.

4. GNODGNOD

Left a contact email that started with kinkykitty@.

5. LILBLUEHAIR

I was helping someone with their resume once who listed their email address as 420bluntbro@…

6. HALF-SQUAT

"Windows 7 was my idea."

7. WANDERSO24

Under "certifications" this guy put "bad ass". He didn't get the job.

8. ANGELICA1883

"I bake great cakes and will share if you give me this job." It was for a mortgage banking position.

9. FAKEMATH

I had a guy put "Cougars" in his interest category. Granted, it was for a bar tending position.

10. MACKDAUG

"My name is Mike and I'd like a job. Here is my phone number. Thanks."

All nicely typed on the first and only line of the page.

11. THEFRIENDLYREDDITOR

My friend used to put "petroleum transfer specialist at British Petroleum" on his resume. In reality, he pumped gas at an BP gas station.

12. BADPAV3

I am an established Aerospace Engineer with a highly technical resume, but at the end of the certifications I put "3CSC Certified" This stands for 3 Cheese Stuffed Crust Certified, as I had to pass a test while a Pizza Distribution Engineer at Pizza Hut. No one has ever asked me about this one item.

13. CJBLAHBLAH

While reviewing resumes that were included for a scholarship application, I happened upon an interesting one, it was a two page resume which I was not too happy about…I turned to the second page to find professionally taken karate action shots of the guy. I felt like an ass for laughing so hard at it because it was a completely serious application, but then that made me laugh harder.

14. BB3RICA

I once got a resume that had been photocopied crooked, and then the top line said: I am seeking a job at Stitches.

Crossed out. In pencil.

15. JUSTA_FLESH_WOUND

I have a buddy who thought it would be a good idea to put his 2.0 GPA on a resume.

16. 2ABYSSINIANS

I was interviewing prospective servers for a restaurant. One young man turned in a resume written entirely in text speak. i.e. Werk Xperince- Bezt bAg Boy in da hizzouse at Da Key FUUD!

I told him the job required he be able to write in English. He took the "resume" back from me and left.

17. DARTHJON

World of Warcraft Guild Leader as an example of leadership skills (listed like a previous job).

18. FREEPASSIONATEME

I manage a coffee shop, and females constantly put glamor shots in the resume. I even had one middle aged woman include a full length picture of herself in an evening gown. It was weird.

19. SEANNZZZIE

Under "Reason for applying with us:" "My parents are rich, and I thought I could live for free off them for a few more years. Turns out I was wrong. Now I need to get a job and move out. I'm lazy though."

20. MORGUEANNA

"Have you ever been convicted of a Felony? If yes, please explain."
Answer: "Yes. Arson. But he deserved it, will discuss in interview."

21. DEFENDPRIVACY

Last employment a girl listed was "exotic dancer at the Doll house" Reason left – Finished degree. They DO exist!

22. GOTPERL

All seriousness: "Italian cuisine logistics engineer". He was a pizza delivery guy. I called him just because he stuck out and was creative. Turned out not to be a fit but got him noticed.

23. ULTIMAPANZER

I had one sent to me that literally said "Why you should hire me: because I'm the shit."

24. ITSJUSTWONNY

"Bachelorette of Science" degree.

25. SAVANTE37

One guy summarized every position with a one-line summary…that sounded like a movie trailer.

The one we laugh about to this day is "a code-slingin' cowboy venturing alone into the Wild West of Java"

The sad thing is his qualifications were really, really, good, but he was just too weird (we did interview him).

26. CATSAILS

"Experience using microwave," on an application to a restaurant.

27. ACCEPTTHEEVIDENCE

I had one once where they had miss spelled 'Union' and written 'European Onion Visa'.

It had me in tears.

28. LEMONWIZARD

In my state's voter pamphlet, one of the candidates in the primary listed "Electro-goth and rap recording artist" in his relevant professional experience section.

To be the governor.

29. CTTOUCH

I always include some photos of me performing karate techniques on the second page of my resume, should

I stop doing this? They are professionally taken shots btw.

30. WHAT_MUSTACHE

I got this. I spoke with a girl and got her resume at a college job fair. She was pretty good, seemed totally legit. I later read her resume, seemed good. On the upper left hand corner there was a small picture of the girl. I've seen that before, it's not a terrible way to remind a reviewer who they spoke to at the fair. Except this picture was of a girl on a college bunk bed in T-shirt and umbros, eating a sandwich. On a resume.

We begged our HR to let us bring her in. We were fascinated. This girl seemed totally normal in person.

To the girl, if you're on reddit. Please tell me if this was a joke.

31. THEOFFICIALPOSTER

"I'm kind of a big deal."

32. GRINR

For a Tech Support position:

"Pickling kitchen – Worked with a prodigious quantity of brine."

He got the job.

33. GLFAHOLIC

Hmm I wonder if I should continue to put "have current Concealed weapon permit" on my resume. My reasoning is that it is a heck of a lot harder to get a concealed permit then pass a job background check…

34. SWINEHERD

"Punctuality: I am almost never late for college or work."

"I am incredibly intelligent and self aware. When I was 12 and my peers were playing with toys, I was teaching myself guitar and reading books. I currently own 94 books."

That 'almost' cracks me up every time I think about it.

35. METALMAN77

"Hobbies include driving around different places."

36. A1MURDERSAUCE

"I have great hygiene and my shots are up to date," on a resume for someone applying for a sales position a few months back.

37. LOVEKARNAGE

Accomplishments: US Citizen.
EDIT: He was born and raised in the US.

38. HAGCEL

Dictaphone. On a resume I got last month.

Expert knowledge of MS word, on a resume so poorly formatted I could barely read it.

The best though was actually in an interview, where at the end the guy tells me in no uncertain terms that there is no way he can pass a drug test. I ask if he has a MMJ proscription an he says "nah, I'm just a stoner."

39. POLKADOT8

A girl had a computer skills section on her résumé, to which she put "no not really."

40. GURRY

Reason for leaving previous job: "They told me I quit."

41. FUBUGER

Had a guy fill out a application and wrote his previous job was "a dick washer at red lobber" more than once. I interviewed him just to make sure it was a dish washer at Red Lobster.

42. TXFLYFLINGER

Under qualifications the guy wrote "Attended Burning Man 2010."

43. BONNSTER

I was a retail manager years ago and I had a young juggalo turn in his resume. His email address started with krystal_meth. Yeah, that one went to the bottom of the pile...

44. TIMMYTIMTIMSHABADU

I had a guy mention that he'd won multiple Dance Dance Revolution tournaments under his "hobbies" section. That was a first.

8

39 People On 'What Is The Craziest Sex Act You've Ever Participated In'

1. [deleted]:

First time I dropped acid I experienced pretty severe synesthesia. Music itself was making me feel like I was orgasming. My wife decided to take advantage of this, laid me down and gave me the most passionate BJ and sex in existence.

What was crazy about this was that as we were having sex I felt that her vagina was a space vortex and her cervix turned into meteors continually hitting my dick. On top of that real squirt was flying everywhere as she was cumming like crazy, but to me it would just hang in mid air and glisten and eventually formed into a giant shining ball of pleasure on the ceiling (yes, it was squirting that far…we have 12' ceilings). I was also narrating everything that was going on and trying to describe what I was feeling the entire time.

We did this twice that night...one of the best nights of my life, though the clean up was intense.

Tl;dr: Fucked a space vortex while being accosted by meteors.

2. dorbin2010:

I was VERY stoned one night at my ex girlfriends house and was crashing on her parents couch when her identical twin came in with her boyfriend and she started giving him head about four feet away from me. The room was pitch black so I didn't think she saw me laying there at all.

I was so out of it that I spent the entire time wondering whether or not she was my own girlfriend cheating on me. When my eyes adjusted, I saw her staring right at me and she winked, smiled, and kept going.

3. TwasIWhoShotJR:

Spanish dark rooms. You show up, shower, and then walk into a completely pitch black room filled with people.

Everything goes.

yeah.

4. wickedsun:

My girlfriend was giving me a blowjob and when I finished she sneezed or something... Came right back out of her nose.

She said it smelled like sperm all day.

5. StayClassynet:

Back in high school my at the time girlfriend wanted to try out handcuffs. So I cuff her to the bed frame and get on with the task at hand. I had pretty good rhythm in my ummm, thrusting. I guess she was pretty into it. With each thrust came a "waeh"-type moan. For no particular reason, I picked up my pace and was giving it to her pretty hard. Her moans picked up in pace too. Now the moans sounded like "waeh-ooo". Basically, it sounded like I was having sex with a fire truck.

So what the hell is a guy to do? I kept thinking about fire trucks. The more I thought about it, the funnier it became. I was trying not to laugh at her, but it was too funny. I was burying my face in her shoulder trying to hide the laughter (we were missionary position)... but the laughter was too much.

And that was that. She thought I was laughing AT her... pretty much killed the sex that day. She started crying and was really sad. I felt pretty bad about everything. It was at this point I recognized that she probably didn't want to be handcuffed to the bed any longer.

Whatever though, I still think it's funny.

6. [deleted]:

My parents are extremist crazy fundamentalists (and I don't use this term tritely). My girlfriend (now wife) and I weren't allowed to sleep together much less have sex so we would sneak the sex whenever we had a few minutes alone.

Well, one day she was on top, in a skirt, and it

looked like we were just rough housing, which was tentatively allowed. My mother, brother and sister walk in and starting talking to us…we continued. Somehow managed to hide the orgasm as we "play fought with her pinning me down."

7. redditluv:

Lady wanted me to masturbate her with the head of a Barbie doll she brought. She opened a little carry drawer made for Barbies. She had about 6 of em in there. She picked one.

She asked me to 69 with her while I masturbated her with the Barbie head.

8. wanttosnugglewithjew:

It involved halloween, a dumpster behind wendy's and a bun.

9. LouisNZ:

me and a mate spit-roasted a chick. awesome.

10. [deleted]:

Golden showers (didn't really dig it), ass to mouth (twice, both times I checked his dick for shit before swallowing it), and two threesomes (FFM).

I'd love to try DP in a MMF threesome. And I wouldn't mind trying out one of those fucking machines…cuz I wanna fuck a rowbutt.

11. [deleted]:

    One time i had sex with my left hand, she was a virgin so we took it reeeeal slow, about half way through her slutty right hand sister (they look similar but are really opposites) wanted to join in, and since my girl lefty was already tired i let her sis take her place, that was a wild 4 minutes.

12. the_guapo:

    I had a threesome with my 8month pregnant ex girlfriend (not mine) and another girl we had just met.

13. dicksquad:

    I'm a male, I let an ex gf fuck me with a strap on multiple times.

14. rocketpants85:

    I dated this girl for a while. She was really a nasty freak. She just loved to get down with sex all the time. It was like, anytime of day, she was like, "Yeah, let's go! I'm so nasty!" And I'd be nailing her and she'd be like, "Oh, you're nailing me! Cool!"

15. texasjoe:

    ))((

16. firebadmattgood:

Plain old butt sex with a long-term girlfriend. Some of you are making me sheepish about not flying my freak flag high enough.

17. jactation:

Not so much a crazy on my part but once my girlfriend was giving me the most fantastic blowjob ever and just as im about to climax... vomited on my dick (and lap)

18. FancyLaserEyes:

Was with my wife and our two friends, who are also married. We had a small party and took ecstasy. Those of that have taken this know that things can often get frisky. We had all done it before, and there was definitely some inter-couple making out.

This time, things got a little fancy. My wife and I slowly but surely started making out, and clothing started coming off. I recall my friend sitting there watching us, while her husband was smoking a cigarette. Eventually, he came back in the room and started working on her. Within ten minutes, they were on the bed next to us, and there was all kinds of romping going on.

It didn't get too crazy—I made out with my friend's wife, he made out with mine, and the girls made out with each other. But there was definitely a lot of fooling around and sex—it was just kept in the respective marriages.

19. rgnysp0333:

Got a blowjob…in a box….in the common room of my freshman year dorm.

Basically there was an end of year charity drive (with no donations at this point) in the common room, which was empty, so we were messing around in the giant cardboard box.

20. Entraya:

At an awesome party for my best friends brother and at about 2:00am I retire to my friends bedroom with her and her boyfriend. We are just chilling and chatting and then her BF pulls out some Oxycontin, we snort it and start drinking. I had already had a cone so I was mellow as fuck and my friend gets down to her underwear to be more comfy. I get down to a T-shirt and panties and we just lay on the bed high as fuck. Then her BF starts to go down on her, at first she's tells him to stop and that its awkward coz I'm there, then she just goes with it. Now my friend is hot as fuck and I'm Bi so I love checking her out and she knows that if she ever wanted to I would be there in a flash and her BF knows this also. After she comes which was like sweet sweet music to my ears , she takes off her top and her BF takes my hand and puts it on her tits. When she doesn't object I get my grope on. Soon after that he literally picks her up and hands her to me, pussy first (she only weighs about 45kgs, naturally always been thin) He then goes behind me and pushes my head onto her pussy and I eat her out. He then starts spanking my ass and leaning over and grabbing her tits. She comes again and he starts to finger her, and she tells me I should get to myself and we come together. I don't really remember much after doing a line off her tits and then woke up at 5:00pm that day with her BF gone home and we didn't speak about it ever again. The weirdest thing is she is so closed off about anything sexual normally and even though she is my best friend in the whole world (finish each others sentences etc) I never knew much about her private life that way, I went from nothing to face in pussy.

21. zmmar007:

    May not be "crazy" but it was certainly the most disgusting for me. I was with this girl I was dating at the time and unfortunately on this one particular night she happened to be on her period. Now we figured, "that's ok no biggy". There was no reason why I still wouldn't be able to enjoy myself, courtesy of a little bit of fellatio right? But my girlfriend decides that a plain old blowjob wasn't enough and she wants to have some more fun with it. She proposed the idea of tying my hands to the bed posts so that she could "tease me" and draw out the pleasure, bringing me to the absolute brink and then holding back. Now naturally, I was all for it, so after whipping up a few knots we were good to go. Now at first things were going great. The blowjob was lasting ages as I was unable to take control and make me reach my desired ending, which of course made the experience more enjoyable. But at some point my girlfriend decides that it's unfair for me to get all the attention. So with me tied up and helpless, she hops up and starts moving towards my face. I sense what's coming and begin shaking my head, just saying repeatedly "No...no, no, no". She soon shut me up as she sat down on my face, smearing her bloody vagina across my lips and trying to force me to go down on her. It was disgusting and I had to sit there, unable to break free, as I clenched my teeth and tightened my lips together with blood trickling down the side of my mouth. It was the worst sexual experience I have had.

22. [deleted]:

Incest, repeated, fully consensual.
   What do I win?

23. undieminer:

Had a girl who wanted me to put it in her ear. Every time I put it near her mouth she'd turn away.

24. FearTheGinger:

M-M-F threesome (I am the F) while on shrooms. Just the one time.

25. Schmibitar:

I had sex in a tree once. It was… weird.

26. biggbuckz:

Everything pretty normal… but in an elevator

27. philter451:

I used to work for a WISP so I had access to a lot of roof-tops which made for some sweet views. I met this really hot girl at a club downtown and thought I might get lucky if I invited her up for an epic view of the city.

Long story short I had sex on the edge of a skyscraper. Most adrenaline fueled sex of my life. I was sitting on the edge with my back to oblivion while she went down on me. At one point she paused and said look down and for some fucking stupid reason I did. Then she got on top and rode me. So fucked up but I still feel like a badass for doing it.

28. bunsofcheese:

The Blue Jays won their second world series (in a row) and I ended up in a parking lot with five other guys and we all had sex. It was kinda fun.

29. gweezer:

Not an 'insane fetish' or anything, but pretty logically insane: I had a threesome with my husband and his ex who he had previously left me for and who I held a hatred/jealousy complex for. It's hard to explain why I decided to do it, so I wont bore you guys with the details-but it was ultimately my decision.

30. [deleted]:

My friend and I were under a blanket in her bedroom.

She stimulated me with a riding crop while sucking my nipples. Meanwhile, her boyfriend was oblivious painting the next room and kept making small talk with us.

31. [deleted]:

I generally think that people are too uptight about sex, and that too few people with reasonably open sex attitudes are willing to speak out.

My most extraordinary must have been a tested bareback party with 5-6 guys and 2 women that I put together a couple years ago. These were hand picked, good looking people who flew in for the occasion, and everyone was recently tested. The girls had a lot of fun, and we, of course, had fun with them. About 2 guys per girl seems optimal to keep the girls busy. There being more than one girl allows for intermittent girl on girl action, which is hot, and relieves each girl of the pressure where she would feel like she has to satisfy all the guys.

Awesome experience, would recommend. It was frustrating to organize – you have to find attractive, responsible, moneyed people who understand testing as a viable method of STD risk management, who are interested in group sex, willing to travel for it, and are available at the same time. It was worth it for the experience, though.

32. sns_abdl:

Wet a diaper while I was going down on my girlfriend.

33. [deleted]:

   I dommed (BDSM) a guy once… strange experience. He was really into me stepping on him in bare feet, his face specifically. He bought a pair of my old shoes that I was gonna throw out for $20.

34. MatthewX5000:

   Rape fantasy.

35. joyfield:

   Me and an ex went to a swingers club and ending up having sex with a lot of people watching. Not really my cup of tea but now i have at least done it and can "brag" about if to at least a few friends haha. Nowadays i am forever alone and the craziest sex i have is with my hand.

36. 3DPK:

   Front seat of a Civic in the mall parking lot during the middle of the day. People walking by and what not. There was an AX commercial shortly after that had a girl with a steering wheel imprint on her back, we laughed for days.

37. FrankManic:

   Once I had someone tied up, more or less at my mercy, and pretty much entirely clothed. I picked them up and threw them in the shower. It was hilarious and awesome and quite kink.

38. historyboy:

Had a dinner party at my place. Two guys three girls. Got blasted off of whiskey and 4loko, then decided to play a strip game. Everyone ended up naked. We then proceeded to do body shots off of each other. Then we got sticky and decided to shower together, before moving to my roommates' room where we became a mass of humanity, sucking, licking, fingering, fondling and fucking.

39. FishboneMinus:

Consensual sex in the missionary position. Shit was absurd.

9

50 People On 'The Secret My Company Doesn't Want The Public To Know'

The following entries have been collected from users on Reddit about their respective workplaces... Thought Catalog has not verified the veracity of their claims, but has collected the more interesting points of the conversation happening over on Reddit.

---

1. [deleted]:

    The recycle bins at Seaworld also just get emptied into the dumpster. They are just there to make the visitors feel better.

2. jrfish:

    I used to work at Claires and The Icing (same company). We got about 15 minutes of training before we were allowed to pierce kids' ears. If people bled on the ear piercing guns, we would simply wipe them off with a tissue, and use them again on the next person.

3. OutofStep:

I worked at a beverage plant years ago that made Arizona Iced Tea, Tropicana, Nantucket Nectars, etc. There was one drink that we produced and the label said, "Made with Spring Water" and it was. Each 600 gallon batch that we made had exactly 1 gallon of spring water poured into the tank.

4. Reed_Himself:

The old hotel I worked at wouldn't change the sheets if they weren't "dirty."

5. TheHosemaster:

While working at HHGregg, customers were told we'd recycle their old TV's for them. Really we just threw them in the dumpster. Can't speak for HHGregg corporation as a whole, but at my store this was the definitely the case.

McAllister's Famous Iced Tea is really just Lipton with a shit ton of sugar. They even have a trademark for the "Famous Iced Tea." There website says, "We can't give you the recipe, that's our secret." The secrets out, Lipton + Sugar = Trademarked Famous Iced Tea.

6. Ingmundar:

My girlfriend works at a small chain of coffee shops here, and was recently told that she had to take down the tips box. She asked why and her boss said it was because the tip box diverts money from breast cancer donations box. The kicker? Her boss pockets the money from the donations box.

7. beefjerkybandit:

If you're ever looking to buy something from a Cash America Pawn store you can tell the how long the items been out and how much the pawn shop has into the item by reading the tag. The date is always printed on the tag. You get a better deal on items that have been on the floor longer. They use a code so an employee can decide how much they can take off if the customer asks. So it goes like this, MARY LOUISE=M-1, A-2, R-3, Y-4 and so on till you get to E-0. So the code will read, YLEE or $45.00.

8. wooddawg33:

I used to work for Sears. The week before our Black Friday sale, we had to mark everything in the store up. I specifically remember marking the treadmills up an extra $500. Then for Black Friday, we marked them back down about $200. They were "on sale" for an extra $300 than they normally would have been.

9. tommishimmy:

I worked at an Orange Julius for 4 years. The fresh

orange juice you guys pay 6 dollars for is half fresh orange juice and half concentrated orange juice.

10. MrKrampus:

Baskin Robbins. Not a huge secret, actually a bit of a plus. But you get a lot more ice cream than you should. I scooped practice scoops for training and the weight it's supposed to be is so tiny. General rule was when the manager isn't there to just make sure the customer doesn't see the bottom of the cup.

11. [deleted]:

I worked at an IVF lab (fertility center) in a major city in the United States. Our center was on the medium-large end, doing about 2000 IVF cycles per year. They wouldn't want the public to know this: exactly what everyone fears will happen, happened. More than once. An embryo belonging to one patient was transferred to a completely different patient's uterus. You hear about this in the news occasionally, but for every case that is published, there are a few that don't go public, and just quietly settle with the patient.

12. sneakystratus:

That Goodwill (in the Willamette Valley in Oregon, company name is Goodwill Industries of the Columbia Willamette) makes ~$100 million in revenue per year (whilst claiming to be nonprofit), the entire board of directors is comprised of business moguls who are either in, or are close to being members of the 1% (just google Michael Miller). Most of the things you donate (80% or above, easily) are thrown away or are sold in bulk as scrap to third world countries. The rest of it is sold at near-new prices ($8 for a shitty shirt in a thrift shop sounds ridiculous to me, and that's barely the tip of the iceberg).

They treat their employees terribly, put forth anti-union propaganda, and (in my store, anyway) use bullying, intimidation, and coercion to keep the employees in line. I've witnessed active discrimination (firing a girl the day she announced her pregnancy; knowing she would be too poor to sue for anything), sexism, racism, and sexual harassment from the assistant manager AND manager of the store I worked in. This was all reported, and nothing was ever done. If there was a person (male or female) that the AM or Manager found attractive or disliked for whatever personal reasons they chose, that person would be either gawked at by the upper staff or derided by them [respectively] in the office. That person would then either find themselves recieving preferential treatment or being given the worst jobs and/or maybe fired [again respectively].

They turn a HUGE profit from donated goods, and provide little actual good to the community. They say they provide employment benefits to unhirable people (those with special needs); what they actually do are utilize government loopholes allowing them to put special needs people to work for sub-minimum-wage, in situations where they don't have any choice (living in group homes or care centres which they have deals with). They end up making ~$5 a day. I wish I was kidding about this, but I'm not. The only other tangible benefit I can actually see from Goodwill are free English classes, which they provide in Salem.

As for their 'standard' employees, they are given a 'competitive wage' of minimum wage, and most aren't given full-time; they are given the 27 hours required to keep them 'part time' so they are not eligible for benefits.

Myself, I was scheduled 40 hour weeks while being on a 27 hour per week agreement. I worked for 40 hours for six months, and yet was ineligible for health insurance. On this wage, I could barely make rent; I had $40 take home at the end of the month (after rent, and before bills), and I survived because of food stamps.

Oh, and when they do take on a full-time employee, they tend to find a reason to fire them just shy of the date when they'd be eligible for benefits. Yes, I saw this happen, to a single mother of two, no less.

Do not give them your things. They are the lowest of the low.

13. virak_john:

Worked at a restaurant — The Elephant Bar in Columbus, Ohio, in case anyone is wondering — and the managers made us save uneaten food from customers' plates and serve to other customers.

If your side of carrots was untouched, it would go back into the warmer and get re-served. Same with dinner rolls and anything that could be passed off as new.

14. tdames:

I had an internship at one of Merck's production plants over the summer. Six months prior, a group of chemists were discovered in one of the abandoned facilities cooking meth. They had a pharmaceutical grade lab, hidden behind a false wall, and were churning out something like six figures weekly. Not something you hear from their PR department.

15. VixenSprouts:

Bank of America... Tellers are all about sales. It is highly unlikely that any of the products they advise you to sign up for are good for your financial situation. Many times they will actually be detrimental, but the position is a sales position, not just a friendly face to help you with transactions.

16. VixenSprouts:

Capital One is famous for a practice similar to this. When you want to close and payoff a credit card, they will not tell you the payoff amount on a credit card you want to close, just whatever the current balance is. You think you have paid the card off but there is now a $0.12 balance, which starts picking up late fees and non-payment penalties because they STOP sending you statements because you think they account is closed. Next time you heard about it when some b.s. law firm is calling to collect on over $400.

17. [deleted]:

    I work in a shipping company warehouse.
    Fragile stickers don't do anything for your package until it's in the couriers hands – maybe.
    Your shipped items are going to get BEAT the fuck up. Wrap it 5x in bubble wrap. If you think you're being too cautious – you're not. Warehouse workers don't care. Your packages are going to be loaded into a hauling truck, stacked in no specific order, slammed around while transported, then throwing around by workers sorting them.
    I'm sure this is already common knowledge. Just a friendly reminder before the holiday season comes full swing.

18. power_overwelming:

    Told to use moldy pepperoni at the little ceasers i used to work at.

19. aerospacemonkey:

    Home Depot has recycling bins out in front of the building, but everything ends up in the same dumpster at the end of the day.

20. lowlikecousteau:

I worked at a Denny's. One of my supervisor's children was conceived in the walk in cooler.

21. luckybone:

at Kansas State University, the FBI has equipment to listen in on all phone calls and data on the network.

22. Cordtus:

When I worked at Wal mart we would throw literally EVERYTHING in the garbage compactor.

Everything includes car batteries, bleach and various household chemicals, large amounts of meat, TVs and other electronics, anything and everything that is easily recyclable.

23. regularITdude:

The Staples equivalent to geeksquad, "easytech", just runs malwarebytes freeware on your computer and charges you a bazillion dollars for virus removal.

24. oddmanout:

I work as a web developer. One of my most disappointing clients was a used car company (one of those ones where they put a GPS in your car).

Basically, they had balloon payments. They'd start off paying something super cheap, $50 a month or something. Each month it goes up by a certain amount. If your notes got to be too expensive, you could actually come trade in that car and get a new one with lower notes, you'd just be starting your payments over for another 6 years.

I had to write an application where the salesman would put in the customers income and expenses, and it would find the "sweet spot" of where the car would be too expensive. We had to make sure that sweet spot was after the payoff point plus a certain rate. That sweet spot was where the company would make the most profit by ripping the customer off the most. It was a combination of how much they paid in to how much the car was still worth. We could predict when a person couldn't afford the car within a 3 month window.

I felt horrible about it, because I knew every one of the customers was going to get ripped the fuck off with it.

25. TheHillSideStoner:

Hewlett Packard barely exists, they've outsourced everything, production, support, R&D, development, management, they probably outsourced themselves by now.

26. mau5trapNB89:

I work for a parking company. Just a little tip: If you get a parking ticket from a private parking agency, you don't have to pay that shit. None of it gets reported.

27. squeakbot:

I worked for Blockbuster. Those collection notices you would get in the mail? Yeah, those were fake. We didn't have a collections agency. They were just meant to scare you and didn't hurt your credit.

28. rubythursday00:

I work at a Starbucks in an airport. It's a licensed store operated by a company called HMS Host. If an employee calls in sick or goes home early from a shift for any reason, even with a doctor's note, he or she receives points that count toward being fired (at 15 points, you're fired).

One day, one of my coworkers told our manager that she was feeling very ill and needed to go home. The manager believed she was sick and told her that she could go home, but she'd get two points. My coworker did not want the points, tried to tough out her shift, ended up passing out, hitting her head on a counter, getting carried out of the store by paramedics, and received the 2 points anyway.

This type of policy results in my coworkers and I coming to work with communicable diseases because we don't want to get penalized or fired. As a customer, I wouldn't want a side of pink eye, diarrhea, or strep throat with a fever along with my latte, and as a co-worker, I don't want to catch something at work from sick people!

In the past month alone I've worked with a girl who had strep throat and a fever so high she was hallucinating, and another girl threw up twice in the back of the unit because she had the stomach flu with a fever.

Strep throat girl called in sick but was asked to come in because we were short handed, and stomach flu girl came in because she had too many points to call in sick.

Does this sound gross to anyone else?

29. KayaXiali:

I was a social worker at an institution for the severely disabled and we had a non-verbal, mentally retarded patient turn up pregnant. Tests were done, establishing paternity. It turned out not only had she been raped by an employee, impregnating her- she also tested positive for syphilis, which the baby's father did not have. So, basically, a severely disabled woman was raped a minimum of twice by two different people while in our care. Fucking shameful.

30. BakoBitz:

I work for a university where, in the 70s, a professor who had risen very quickly through the ranks to become chair of the Faculty Senate, took a group of students to Egypt and abandoned them. He took off with all the money that had been designated for the trip and left them with nothing but a huge hotel bill that no one could pay.

Last year, when his biography came out, we discovered the professor was Steve Jobs' biological father.

Also, our most famous grad? Ted Bundy.

31. WunderOwl:

The New England Patriots make hourly front office workers clock out early and keep working so that they won't have to pay for health insurance.

32. getouttatownguy:

I worked at Sbarro's Italian Pizza for a year, and I was the only one to pull off the counter top (where we actually make the food from scratch) to clean underneath. I found an infestation of maggots that had to have been getting into the food we made everyday for god only knows how long.

33. Series_of_Accidents:

Don't ever trust market research. Just don't. Cunningham Field and Research Services is one of the many companies that collect responses for companies like Nielsen, etc. And… they totally falsify the fuck out of their data. Need a 90 year old woman for your quota? Well why not use a 15 year old boy and tell him not to tell anyone. Or… if it's really down to the wire, why not enter it yourself and have the check made out to a friend? I had to quit, the lack of ethics was killing me.

34. Hermwad101:

The restaurant where i work is well known for their home-made garlic toast. It comes with every entree. If the table doesn't finish their basket of toast, whatever is left in the basket goes back in the warming drawer…. And then back out to another table. I always try to encourage people to take it home with them, usually saying something like, "you might as well take it home, we're just going to throw it away and it'll go to waste!" just so I don't have to put it back in the drawer. I can't help but think about that one piece of garlic toast that keeps traveling from table to table and never getting eaten and continuously gets recycled. Eew.

35. [deleted]:

Ancestry.com would destroy books they had digitized because they didn't have the storage space. That place infuriated me.

36. ne1av1cr:

United States Military.
   You would be amazed how many mission critical decisions are made with rock/scissors/paper.

37. ayb:

When the potential investors came to visit our startup office, we got our friends who didn't work there to come in and sit at computers and pretend like they were working so it looked like we had more employees.

38. The_Drama_Rich:

I worked at Best Buy when the Xbox was pretty new. At Christmas, I was forced I "bundle" the system together with crappy games and controllers and other useless items. We would then only sell them the bundle and hide the single system in the back and tell people the bundle was the only ones left. Most people didn't care.

The worst was when a single mom came in with her 2 kids. They were super excited to get an Xbox. I could tell she didn't have much money and this was a big deal. They were all devastated when I tried to give them the $650 bundle. Knowing I would get in trouble, I told them "let me check one more time in back". I happened to "find" one just for them. I took the heat when they tried to check out, but I gave no fucks.

39. C_IsForCookie:

At Best Buy when we recycle old products, we can take them home. I have a friend who works for Geek Squad and has a shit load of Macbook Pros that he convinced people were broken and unfixable so they'd "recycle" them. He takes them home, fixes them, and keeps them.

40. My_fifth_account:

Cabela's – Their shipping charges are based on item cost. You can buy a $100 deer stand that weighs 90 Lbs and it'll cost you less to ship than a 2 Lb camera. Also, keep your receipt as you can return anything there at any time because the managers are so scared to lose their jobs. You can go buy something for the hunting season, go use it, then return it months later just by saying it didn't live up to your expectations. If they balk, threaten to complain to corporate and they'll do it. If not, complain to corporate and then they'll do it.

ETA: Goes without saying to stay away from their credit card, but every person they get to sign up for it Visa gives Cabela's $300 pure profit, at least that's what the amount was when I worked there years ago. They care more about signing people up for that credit card than selling you anything else in that store because it's 100% profit which is why you get attacked when you walk in the door and by every employee you talk to. Easiest way to avoid the conversation? Just tell them you already have one and all pitch lines stop.

41. sadtastic:

Back in the 90s, my boss hired an ex-con because the state was offering some sort of tax incentive to hire people out of prison. The ex-con was eventually given keys to one of the buildings. He took two girls there after hours and murdered them.

42. T-Luv:

One time, working at Domino's, this lady on the phone pissed off my coworker, so he put a fly in a black olive and put it on her pizza. I told him not to, but he did it anyway.

43. It_does_get_in:

We incorrectly installed the seat safety bracket mounts

in over 750,000 cars, and there has been/will be no recall.

44. hex498:

TelemeriCorp calls clients on the "Do not call list."

45. JustForCancer:

Cable Internet Provider: No matter what speed of internet you have it costs the company the same amount.

46. Riverbed19:

I worked for a sister company of Hormel. We made meatballs and pizza topping meats. We were supposed to, before and after the almost 3 years I worked for them, clean our equipment between species. So wash all the equipment after running beef, and before pork. They didn't start doing that until I left. So anyone who couldn't have certain meats for whatever the reason may be, if you ate certain pizza's or meatballs around the Midwest, you probably had a little bit of everything in there. (We still washed out for allergens! Soy/Wheat/Dairy/etc.)

47. annamarieraven:

When I worked for Chase mortgage collections department, during the time mortgage modifications were being rolled out by the Obama administration, our customers were constantly asked to call back to check the status of their mod, lost paperwork, etc and told to pay less than their full mortgage payment so that they would be declined and then we would foreclose- even though we knew they would be declined from the start because the bank made more money foreclosing than modifying peoples' mortgages. This was about 3 years ago.

48. Bankerofblood1:

The lobster you picked from the live lobster tank is hanging out in the walk-in till you leave. The one you're eating was frozen.

49. [deleted]:

I worked at a Fish Market and 90% of sushi restaurants are lying about which fish they are using.....

50. seifermaster:

Wal-mart made me destroy around 30 brand new bicycles because it was the end of the summer and they couldn't sell them. Apparently they don't want people digging in the trash for a new bike, so they made me and another guy take a sledgehammer each and destroy them as much as we could. They wouldn't give the bikes to charity because they could apparently come back against them if the bikes were defective, which is complete bullshit since the bikes were new.

10

21 Pieces Of Life Advice From People Over 60

The inquisitive souls over at Reddit are always looking for some good advice from any source they can, and this week they decided to ask all of the Redditors in the over-60 crowd if they had any pertinent words of wisdom to pass down. The response was enormous, and some of their thoughts are the kinds of things that we need to print out and tape up to our bathroom mirror right away.

1. DAVELOG

    Stuff is just stuff. Hoard time instead.

2. MAMA146

    Choose your mate with your brains as well as your hormones. Be picky.

    If you're getting overwhelmed just return to the immediate present moment and savor all that is beautiful and comforting.

    Your Life is not as serious as you think it is.

3. ELECTRONICAT

Knees are important. cherish them

4. AABBCCATX

I asked my grandpa something like this as he was dying. So me, no I am not 60 I am 28 but he was 83. My grandpa told me three things

"If I knew I was gonna live this long, I would have taken better care of myself."

"The right job is the job you love some days and can tolerate most days and still pays the bills. Almost nobody has a job they love every day."

"My family is the only thing I care about anymore. Remember that [aabbccatx]"

5. DEMO7

I'm 62. Most of the advice here is good but I would emphasize two things. Take care of your health and your finances. Start eating better and exercising regularly. If you put on weight now, it will be much harder to lose it later. And if you get into the habit of eating a very high calorie diet, you'll probably continue that diet as your metabolism slows down and you'll put on the weight then.

Don't fall into the credit trap. Live within your means. I know two kinds of people; those who save for things that they want and then pay cash, and those who buy on credit and pay interest on top of the purchase price. Once you start doing this it can quickly become impossible to change the pattern. All of your extra income goes to paying the credit cards/car payment/etc. and you can no longer put any money in savings. Then when you need (or want) to buy something, you have to do it with credit. Add a sudden large expense such as a medical bill, and you may never escape the pattern.

6. MONEDULA

Well, as the old saying goes: a stitch in time saves nine.

Those weeds in the front garden? Pull them out now, or in a few months there'll be ten times as many of them and they'll be five times as tall. And next year they'll be bushes with roots that are a real pain to get out. (Personal experience)

The flaking paint on that window-frame? Paint it now, or it'll rot and be far more work to fix. (Personal experience).

That nasty sound when the car-wheel hits a bump? If you don't fix it and the suspension breaks when you hit a pot-hole on the motorway, you'll be stranded there with a forty-tonner bearing down on you. (Fortunately not personal experience – I'd learned something by then.)

7. CIPHERDEXES

Books. Read them. 64 here, and your mom. All the clichés apply (sunblock, flossing, travel). But don't stop reading books, lots and lots and lots of books. Crappy ones, disturbing ones, difficult ones, fun ones. You can only live your one tiny life, but with books, you can live thousands more.

8. WIZARD10000

I'm not quite 60 but am closer to 60 than 50 so I'm gonna take a shot anyway.

The most important person in your life is the person who agreed to share their life with you. Treat them as such.

Children grow up way too fast. Make the most of the time you have with them.

A friend will come running if you call them at 2am; everyone else is an acquiaintance.

Your job provides the means to do what's really important in life, nothing more. Do the job but live for your family.

9. ZAZZLEKDAZZLE

"Your job provides the means to do what's really important in life, nothing more. Do the job but live for your family."

I would say live by the spirit of this concept but not necessarily the letter of the law here. If we all treated our jobs like 9-to-5 money-making nuisances that interrupted our real lives, we would have no vaccines or antibiotics, no expert surgeons who get up out of bed in the middle of the night to reset your knee so you will be good as new rather than crippled for life, Dr. Martin Luther King Jr would probably never have planned the march on Washington, and Freud would have stayed just a regular doctor and rather than founding psychology as we know it today.

The answer is not necessarily to become detached

from your job, but if you love your work and it's important to you, involve your family. The two do not need to be separate. If you do something cool your kids are going to think it is insanely cool. Bring them in on their days off from school, if you bring work home talk to them about it, if you have to go in on the weekends bring them for a while and give them play "work" to do — then take them out for a treat. The greatest memories of my childhood were going into the lab with my physician/scientist father. I was inspired by how much he loved his work and his dedication to saving lives and finding cures. And I freakin' LOVED working those machines in the lab and looking at slides on the microscope. Plus, this let me know that I was more important than his work.

My father's work was demanding, he worked 10-11 hours a day, and often on weekends and holidays, but he was NOT a workaholic. He was always home in time to help me with my homework (when I was older) or give me my bath and read me bedtime stories and have a bedtime heart-to-heart (when I was younger). You don't have to do something obviously cool like science to involve your kids, your kids will think anything you do is cool, the trick to make them feel involved and use work time as bonding time with them as well. A friend of mine who is a lawyer often has his son into the office. He gives the kid his own computer and legal pad and he does just what daddy does (although it involves playing games on the computer rather than writing briefs, and drawing pictures rather than taking notes, but you get the picture).

It's not just about kids either, make sure your partner knows he/she is valued as well. My partner and I often "telecommute" from each other's offices, or work together in the house or at cafes, often taking snack and conversation breaks. It is wonderful quality time for two introverts ?

I think it's OK to devote a lot of the time in your life to your work, just don't make it separate from your actual life. I would say: work a lot if you want but make it part of your life, and don't be a workaholic — if Barak Obama can make the time to have a nice family dinner every single day (as long as he's not traveling for work), you can have work you love and are dedicated to and spend great time with your family.

10. DUCKDELICIOUS

I'm exactly 60 and agree with what's been said so far. Especially how fast your children grow up. It's why grandchildren are so wonderful…I realize it now and appreciate and savor every single minute with them.

I would also say that years go by in the blink of an eye. Don't marry young. Live your life. Go places. Do things. If you have the means or not. Pack a bag and go wherever you can afford to go. While you have no dependents, don't buy stuff. Any stuff. See the world. Look through travel magazines and pick a spot. GO!

If you have a dream of being or doing something that seems impossible, try for it anyway. It will only become more impossible as you age and become responsible for other people.

We have one time on this earth. Don't wake up and

realize that you are 60 years old and haven't done the things you dreamed about.

11. PHFAN

As a 60 year old, I'd say to 30 year olds that you should date someone twice your age.

12. RAREAS

When you meet someone for the first time, realize that you know nothing about them. You see race, gender, age, clothes. Forget it. You know nothing. Those biased assumptions that pop into your head because of the way your brain likes categories, are limiting your life, and others' lives.

13. GRADUALJEWISHMOTHER

I would say to appreciate the small things and to be present in the moment. What do I mean? Well, it seems today like younger people are all about immediate gratification. Instead, why not appreciate every small moment? We don't get to stay on this crazy/wonderful planet forever and the greatest pleasure can be found in the most mundane of activities. Instead of sending a text, pick up the phone and call someone. Call your mother. Really, call me. I know you are busy and have a new girlfriend and an important job, but please, you just have one mother. I promise I won't ask about children or about your shiksa girlfriend. No, really, as long as you two love each other. Can she cook? I meant to tell you that you looked too thin when I saw you last week.

14. MRRIZZLE

I'm here with my grandma, she says: "Fuck Bitches, get money. Yolo."
I'm just kidding, she said: "what is read it (reddit)? I don't have that, I have hotmail."
Love you Grandma.

15. PATSAYS1

Not quite 60 but getting there fast. If you are a US citizen chances are you've lived a pretty easy life even if it didn't feel that way. In a world of abundance here's what I've learned. #1. Either have a great partner or don't have one at all. My spouse has made everything better, and I put her on a pedestal. When nothing seems to be going right, if you've got each other, you've got everything. Any time I doubted myself she believed in me more than I did. Thought #2 – Take risks! we think in terms of black and white, the world is gray. It's really hard to "lose everything" unless you die. If you get fired there's another job, if you go broke, you can rebuild. If you don't take risks, (smart risks) you'll look back and say "I wish I had." #3 Blood is thicker than water. Have kids if you can, they are worth the trouble. Protect your family, expect them to protect you. #4 Your brain thinks you're 18, your body gets old. Don't be afraid of going to the doctor, and spend whatever you have on your health. Your spouse, children, family, colleagues need you #5. Keep your word, even though most people will end up disappointing you with theirs. #6 Learn to be generous, it feels good. #7 Never stop learning, seeking more knowledge, – especially now, what you thought would always be a career will be obsolete. If you are planning a career now – do something related to energy, water, or waste and you'll probably have a job for the rest of your life. I'm sure there are more, but that's a good start.

16. JERRYVO

Marriage is not 50:50. It is 100:100. Remember that when you wake up and when you go to bed. Every day.

17. SHURIKANE

I was taught a lesson that seemed shallow at first but it ended up being very wise (even if the one who said it wasn't): "Work not too fast, and not too slow."

This was on my first day of a physical warehouse job. I was a little surprised at this advice, being all young and naive and wanting to give it my 110%. I thought at first he was gaming the system.

On that day, around 2:00 PM, I was out of breath and my body wanted to kill me.

I learned the virtue of sustained work. Pacing oneself. Adopt a steady rhythm so that you're not completely destroyed at the end of the day. This was especially true at this job where the summer season saw us doing 80+ hour weeks of physical work.

18. VECTAUR

To your last point…a friend once told me, when I was working ridiculous hours and was stressed beyond reason, something that has really stuck with me. He said:

"Nobody ever dies wishing they had worked more."

Since that very day, I have tried to get out of work on a more reasonable schedule, saving the crazy crunch hours for when there truly is a crisis at work. I still think it's important to excel at work, as getting laid off would of course, suck, but I just try to stay very focused and get it done in as few hours as possible. I spend more time with my wife and doing hobbies as a result.

19. TFSD

I'm 60, and I generally agree with the other answers, but I want to give a slightly different perspective. Remember that life is like a bank account: You don't want to spend everything you have right away, but you don't want to be a miser and save every penny. Yes, you're only young once, but, with any luck, you'll also be old at some point. Plan on a career, but don't let it overcome the rest of your life. Take care of yourself, but don't make it an obsession. Focus on your kids, but leave room in your life for yourself. Save enough money so that you'll have enough for the future and for emergencies, but spend enough now to avoid looking back with regret. I've tried to balance living for the moment with planning for the future, and it's worked out well. Many of my friends who ignored the future when they were in their 20s and early 30s now bitterly regret it, but many of my friends who did nothing but work and strive during the same period say they have the same level of regret.

20. SCREAMWITHME

Collect experiences. I don't have many regrets, but I do wish I would have travelled more when I was younger. And this: When you have kids you will be time warping. Spend as much time with your children as you can. Turn off the TV, get away from the computer. You will never understand the impact you have on their lives. Make the most of it.

21. AESU

Life isn't serious. Success or failure mean nothing in the scheme of life's existence, never mind your life. If you are fortunate enough to be born into relative wealth, enjoy it. And if you aren't, do everything in your power to make everyone's lives better. The only goal we can ever really have as a species is equality of happiness.

11

50 People On 'The Secret I Am Terrified To Tell'

1. throwaway215091:

   Two and a half years ago I was in dire financial straights, so I sold my home to keep my struggling business afloat. I neglected to tell the owners that they have an 800 sq. ft. bunker on the property that I built about seven years ago. The bunker that I've called home since I sold it. The entrance to it is well-hidden, but I still come and go very early/very late in the day.

   I'm a single man who keeps to himself. I'm now in a situation where I could move somewhere else, but I love this hidden paradise so much.

2. Tomgoldaccount:

   I cut off all contact with everyone I know and moved to Kenya, I tell people a fake name and a fake background and have made it appear to my family that I died on boat trip in the Pacific. No I am not joking. I am dead in the United States.

3. iGotYouThisCake:

I run a cake business. I charge people hundreds for wedding cakes... Every last one is made using Pilsbury cake mix I buy for $1 a box at Walmart. I suck at baking. Every time I've ever tried to make a cake from scratch it sucked. But baking is like.. My whole deal. My friends all call me the cake girl. It's like my whole life is a lie. People compliment my cakes all the time. Telling me how delicious they are. Telling me it's so much better than box mix cake. Telling me they could never bake a cake so delicious. Well guess what? For $1, they too can make a cake just as delicious. Just add oil, eggs and water. In my defense, I love cake decorating. I make all of the frostings and fondant from scratch. I just hate baking fucking cakes!! I base my prices mostly on the decoration of the cakes and not of the cake itself of that makes sense. Still... No one knows about this except my husband. Even my best friends think I fucking slave over the oven mixing and baking these damn cakes. I have been doing this for YEARS. If anyone knew my business and reputation would be in the toilet for sure. :/ I keep telling myself I have to learn how to make the damn cakes without the box mixes, but I never do it. I feel like such a sham sometimes.

4. ThrownAway2389:

I once helped out my a female friend's family by taking care of their cat for a week. Every day for a week, I would go over there and snoop around their house. I found my friend's diary, and proceeded to read the entire thing. I used this information to get her to like me, and she is currently my wife.

5. morningandamazing:

I don't want to be with my girlfriend anymore, but she might have cancer and I feel like I need to stay in the relationship.

6. HalfEducated:

I faked the last two years of college education. My parents put so much pressure on me I couldn't handle it (I was suffering from severe depression and anxiety) so I faked it all. Lied to everyone. Made up fake transcripts. I just got my foot in the door in my desired field thanks to a friend as they hired me as a subordinate. This place only hires college grads but no one double checked my credentials since I was recommended. My hopes is that if I need to find another job I'll have been at this place long enough to get it by experience alone (I work for a very prestigious company). I'm not bad at my job. I'm actually quite good. But my fear is eventually I'll hit a wall and the lie will come to light. No one has known this for the better part of a decade.

It's a relief to finally say it "out loud." I can't even tell those I love. My silence is my prison.

7. therealandrew:

   When I was 17 I had a argument with my father and told him to fuck off, later that evening he hung himself. Our argument was the last time he spoke to anyone in our family and for that I feel a terrible amount of guilt for. Instead of him saying good bye and I love you to my mom and brothers he got told to fuck off before he went and killed himself. My punishment is to live the rest of my days in shame and guilt. He never left a note either.

8. cunt_rocket:

   I used to be a Police/Fire/911 Dispatcher, but had to quit because it nearly made me suicidal. I actually had thoughts, but had to drive 40 miles to go to a center/hospital where no one knew me for help. I have nightmares about a few calls I took where the caller killed themselves, shot someone else, or passed away on the phone with me. To this day, a few years after resigning, I still can't listen to a phone ring, or sirens go off without having a mild panic attack. I am fairly sure it's a form of PTSD, with flashbacks, nightmares, panic attacks, and an inability to function sometimes, but I'm embarrassed and scared to tell my fiance, or go to a doctor for it. I know there are soldiers out there with real PTSD that deserve help far more than me... I am very good at hiding it though. I also sometimes wait until my fiance goes to sleep, and I will then go sit and pretty much cry for several hours. It's hell.

9. Throngsong:

Everyone thinks I have a good job and roommates but I've been homeless and a prostitute for over year.

10. yesthisisthrowaway23:

IT guy here, it's amazing what people will do on their computers and say in their emails despite having to sign a waiver that all computer activity at work is monitored and recorded.

I have half the company's banking, social media and personal email account info and passwords. I know who is secretly banging who at the office behind their spouse's backs. I know who is cybering at work and jerking it in the bathroom almost daily. At least they tell their sex chat partner they're running off to the bathroom to jerk it, haven't felt the need to check the validity of that one. I know when people are having martial problems, financial problems, I even know one person here had their children taken away because a social worker found cocaine in their house. I know who is embezzling money, I know when people get fired for completely bullshit reasons (like they just want to replace them with someone younger and nicer on the eyes), and I know who my boss is buying xanax and vicodins from.

Basically I have a treasure trove of my coworker's secrets. I won't actively do anything with this info, but it's nice knowing I have the ammunition there if something were to ever happen.

11. iamfromcanada:

There was a girl who I had a crush on the moment I saw her on my college campus. She ended up dating a douchebag dude a few weeks later. I happened to end up sitting in a study room with him and a few mutual friends. He talked about how he didn't think she was that attractive and how he liked other girls. I wrote the girl an anonymous email using one of those websites telling her about the things I heard and how the guy was a dick. She ended up breaking up with him after she found out he was cheating.

The girl is now my girlfriend of 6 months. She has no idea (and is sitting across from me in the library). I've never told anyone this before.

12. throwaway3708:

When i was 15 my parent's were going through a divorce, my mom worked night shifts and my dad was living with a friend of his. One night my sister who was 19 at the time came home pretty drunk from a party. She was acting goofy and fell on the couch next to me. She started grabbing my leg and laughing and we started fondling. We ended up having sex right there. When we woke up the next day she had no recollection of the night before so i just kept my mouth shut.

Fast forward to when i'm 18. Sister is home from college and dad is over for a visit. they get into an argument and in a fit of rage my dad announces how he has never forgiven her for the abortion she got when she was 19 and subsequently killing His grand child. (he's very religious)

I then realize the baby she aborted was in fact mine.....and as far as i know, i am the only one who knows since she has never mentioned that night.

13. Homycraz2:

Not me but one of my frat brothers in college knocked a girl up. A month later she had had lost the baby. I was using his phone one night to find my phone when his dad texted him, i swiped the lock causing it to open up the chat thread revealing the messages that explained the story.

The day he found out he drove with some of our other brothers to Mexico and he came back with RU486, the abortion pill. He had apparently spoken to his dad who forced him under threat of pulling him out of college and cutting him off to secretly sneak the girl the abortion pill. I dont know the logistics of how he did it but apparently he secretly poisoned her causing her to lose the baby.

He has no idea I know and I doubt anyone else does.

14. [deleted]:

I have a blind brother. When we were young, I used to get so frustrated at all the extra attention he received and how I had to be more responsible with my sibling than my peers. So, when my brother and I would go play, go to the store, or just generally go anywhere without adults, I would abandon him somewhere unfamiliar to him. Then, I would stand off quietly and watch the anxiety set in as he tried to figure out where he was and what was going on.

15. Amgpu:

    I accidentally killed seven people.
    I put a rag into a new water heater exhaust to keep debris out and installed it in a rental.
    I get a call a week later, there's been an accident. I show up and there's a ton of ems and police. They ask me where the gas shutoff is, and I go down to shut the gas off and see the end of the rag I forgot sticking out of the top of the heater.
    Ripped the rag out, shut the gas off and head upstairs only to be told all the tenants were DEAD.
    I drink all day now and sleep. It's killing me from the inside every single day, but if I say anything my family is ruined; we have a bunch of rental properties and we'd be shut down.

16. britishNIGGA:

    I hate all of my friends. Literally. I don't have anything in common with any of them, and don't care. But I'm too scared to be alone and have no one else to go to so I keep hanging around with them.

17. [deleted]:

    My own secret, is that I'm still deeply in love with my (now married with kids) first love, nothing will ever happen and it is ridiculously hurtful, but w/e, life goes on.

18. omfg_name_taken:

I have memories of my sister (five years older) and I playing a roleplay game when I was younger that I think would be considered sexual abuse/molestation if I told anyone. I don't remember how old we were, but I know she was around the age where her breasts were developing. When home alone we would play a role play game where she was a boss and I was a secretary, and the boss would always sexually harass the secretary. It ended in my sucking on my sister's breasts while she would lie on the couch with her shirt off.

My memory has always been really horrible, so I only remember patches of this, but I remember that it never felt sexual. I don't actually trust my memory enough to feel confident that this really happened.

I love my sister, she's my best friend and I would never want to damage our relationship by ever bringing this up and asking her what really happened. It is a secret I will carry with me and never reveal.

19. rattlesnaker:

I still have "imaginary friends." I'm almost 30.

I lost them for a while. I don't know why or how, but it they were gone. I couldn't see them or hear them any more, not the way I used to when I was younger. It made me was miserable. I kept hoping for a way to get them back.

Two weeks ago, I somehow managed to finally break through whatever the barrier was. I have spent the past two weeks hanging out with, and talking to, a character from a well-known TV show.

I can't really "see" him visually, but I can see him with my mind's eye. He goes almost everywhere with me. He's sitting on my bed right now, waiting for me to get off my computer. (I promised I would get off a little while ago, but I had to check reddit one last time.) He's been coming to work with me every day for the past two weeks. I share my food with him. (I kind of mentally duplicate it for him, since he can't touch it in reality.)

I love it. I'm happy again. I realize most people would say he isn't real, but something about him is. I don't care. He's real to me.

20. thisiscaptainmeow:

I used to masturbate a lot. And when I was 10 I had a technique where I'd let off a load into a sock then wash it and quickly dry it, now I couldn't leave it hanging outside or use a dryer otherwise my family would've seen it and probably smell it or whatnot. So I'd put it inside my gas heater unit. Unfortunately my sock had caught on fire inside the unit, blew it up and set my house on fire. Only my brother was home at the time, and he managed to survive the house did not. For 5 years we stayed from caravan park to caravan park whilst we waited for confirmation that it was not arson and we could receive an insurance payout. We eventually did and scraped together money to start rebuilding the house. The house is still being rebuilt to this day and it shames me anytime I have to visit my parents living in a tiny mobile home where my backyard once was.

21. [deleted]:

I have been pretending to be colorblind to everyone I have ever known, including my own parents since I was in 3rd grade. I am now 28 years old. I even convinced an optometrist of it.

22. a_blackmailer:

When I was 13 I caught my father in bed with my 15 year old brother's girlfriend (also 15). I haven't seen her since, but I've been blackmailing my father with it for the last 6 years.

23. throwaway891872348:

I was hit by a truck a few years back and was diagnosed with retrograde amnesia and awarded a 2.5 million dollar settlement.

I have used this money to move into Florida and I currently live in Boca Raton.

I faked the whole thing because I hated by life and wanted an excuse to leave. I haven't seen my family since and have made a new life.

24. donnybrook00:

my grade 6 teacher let me touch her boobs once.

25. FelineOfTheSea:

After graduating from high school, I went to a small out-of-state college where no one from high school knew me. I was told many times how impressive my false Australian accent was, so I decided it would be great fun to go through college pretending to be from Australia. All of my friends and even my girlfriend of two years think I'm Australian. I have a completely fake Australian identity, family, and past. I will soon be graduating, and I plan on asking the girl to marry me. Everything she knows about me is Australian I don't know how to tell her she doesn't really know me. Guess I'm forever a bloke.

26. [deleted]:

My Great Uncle Jack used to live with my family. One day, he got drunk and had a bad fall that ended up causing him to bleed out, I ended up finding him (I was 14 at the time, and had never seen such an awful sight) and lost consciousness due to all the blood. When I eventually recovered, I called the ambulance and stayed with my uncle, he died in the back of the ambulance, holding my hand. No one knows about what happened to me, and if they did they would realize that I'm the reason he's dead.

27. intsa:

I've never attempted to kill myself, and I doubt I ever will, but I just want to die. I'm an incredibly happy guy odd enough. I truthfully am happy, but whenever I think about getting shot, or getting cancer, I get a little excited. I wish I was one of those deaths on the news, shoot I'd love to take someones place, they want to be here more than me. I'll never actually kill myself even if its just for the sake of others who need me, but I can't stop wishing that someone else would kill me. I'm done being here, I'm done dealing with the crap. I'm just burnt out and I don't want to be here anymore.

28. fayuluire:

Every night when I go to bed, I have a little pillow and assortment of blankets that I pretend is this girl I like. She would never like me in real life (in fact, she doesn't), so I just play pretend. It's inherently creepy but it's what keeps me from being a total wreck all the time.

29. Turtles94:

Last summer, when I was 16, I found out that I was pregnant. I come from an extremely conservative and Christian household, so I was too scared to tell my parents. They also didn't know that I was dating my boyfriend of the time, because he is Hispanic. I decided to get an abortion, but didn't have the money to fund it. My boyfriend had a job, but kept encouraging me to keep the baby. I tried and tried to gather the $300-600 necessary for it, but it was so hard. I ended up having to order RU486 (the abortion pill) from a sketchy website online with my own money, because I was so scared and desperate. I ended up getting really sick from it and had to explain everything to my mom on the way to the hospital. Since I hadn't gone to the doctor before, I wasn't aware of how far along I was. I was over 6 months pregnant, and had hid it from everyone in my life, other than my boyfriend. I hadn't imagined the emotional side effects, or what would happen afterwards. I ended up giving birth to a baby much bigger than I could have even imagined, and he suffocated to death almost immediately. As if the shock of this wasn't enough, the doctor called the police and I got investigated by a homicide detective. I hated myself to the core and still do a few months later. The thing is, that no one would expect this from me. At all. People think of me as such a "goody two shoes" and I was recently voted "class clown." No one could imagine that I had an illegal, late-term abortion at 3 in the morning. No one could even tell that I was 6 months pregnant, because I only gained 6 or 7 pounds. No one would imagine that I'm being investigated by the homicide detectives or that I fight off thoughts of suicide daily.

30. deejaweej:

I was falsely accused of raping a girl in high school. The resulting ostracizing was very scarring, and that is just the tip of the iceberg. I outran the stigma when I left the state for college. If it ever catches up to me like it was in high school, I'd probably become suicidal. How many times can you endure people telling you that you're a monster before you believe them?

31. nottherealjethrotull:

When I was about 12 I went with some family to the family dollar. My mother and cousins went off to go look at generic groceries so I decided I would just spend my time hanging out in the toy aisle, in the toy aisle there would always be these bags of marbles that other kids would open and leave laying there so I decided to fling marbles across the floor and one just happened to reach one of the far off aisles. So about two minutes later I hear a loud crash and someone scream "Somebody help this man!". Being the curious child I was, I ran over to see what the commotion was about and I find everyone gathered around this guy who had seem to have fallen from the ladder as he was getting something off the top shelf. The guy is seizing out and blood is coming from his head as he laid there and his face seemed to be turning blue. My mother whisked me and my cousins away and we left. Next time we went we talked to the front cashier and she said that they called the paramedics but by the time they got there he had died from choking. Apparently when he had the seizure he was choking on his own tongue. The cause for the fall according to the front cashier was that he had put the ladder on a marble and didn't check it before he got on it. When I heard what the cashier said I just stood in disbelief thinking I was going to jail, I tried telling my mother many times but all she did was say that I imagined it.

32. Throwaway36363636:

I'm a 25-year-old female high school teacher. I've gotten myself off on multiple occasions while fantasizing about fucking one of my 16-year-old male students on top of the desk in my classroom.

33. DuncanGilbert:

My mom died when I was 17 and when it comes up I use it to garner attention for myself. In reality, I never met her and she has never meant anything to me other then a name.
    I feel so empty

34. iamaliar22:

I told my entire family I was able to transfer out of community college and into a university, but I never finished up the requirements. So since I live at home, every day instead of going to school I go to the local library and bs. My lies are so extensive, I even go to the campus and meet my girlfriend for lunch sometimes. I've made fake transcripts to show my family, and to make it look like I'm actually studying I go to MIT opencourseware to look up facts that I "learned in class" that day. I have become a remarkable liar. I hope to be transferring in the fall and then I look forward to living a normal life. Coming clean is not an option at this point.

35. Uppgrayyed:

After my mother left my father, he developed a really

inappropriate attachment to me. I was 19 and my brother moved in with his girlfriend. Dad was suicidal, and had no family or friends close by, so I was it. For the first year, he would wake me up at 2am to sit with him every night until he cried himself to sleep. After 4 years of cleaning up after him, making sure he ate, and generally remained alive, I discovered that he had been using the attic access in his closet to sit above my personal bathroom and watch me through a peephole. I wanted to dismiss it as paranoia, but there were too many physical signs that made it reality. Moved out shortly after that because I couldn't bear to look at him. I'm 29 now, and no one in my family has any idea that this ever happened. I know that he was going through a rough patch, but I feel violated and dirty every time I think about it still. I also have huge amounts of guilt because I hate him for putting me through it.

36. throwaway1450:

This will probably never be seen by anyone but fuck it. My father once owned a cat who loved to suck our earlobes for whatever reason. About half a decade ago my father left me alone in his apartment with his cat and I don't know exactly why but I just grabbed the cat, went in the bathroom with it, laid on my back, put it on my chest and let it suck my earlobes while masturbating. I find myself fucking disgusting when I think about it but I still think that it was one of my best faps.

37. RichTraitor:

My dad got rich by associating with a scumbag that has his own construction company. Scumbag bribes city officials to approve unstable skyscrapers that would collapse with a 4.0 earthquake and my dad makes all the paperwork discretely. In exchange, multimillionaire scumbag persuades his other loaded friends to hire my dad as their lawyer.

I'm now trying to get into office in the next 30 years to revert most of what my family has contributed to.

38. moarnames:

I'm a 30 year old woman and I've never had sex or kissed anyone. I've never had a boyfriend or a girlfriend. There's nothing physically wrong with me, nor am I unpleasant to look at. I masturbate a few times a month, mostly because of a biological need rather than actual desire, I guess. I've never fantasised about anyone or felt any physical desire for anyone.

39. ttthrowayyy:

Me and my cousin have been doing it for 10 years now. It started when she was 12 and I was 13. We had to babysit the younger kids in our family while the parents went to a party, and when they fell asleep, me and her got to talking about a lot of stuff. I made a move and started kissing her, and she didn't resist. We ended up doing it on her bed that night. We would end up fucking almost every weekend when we lived with our parents, telling our parents we were going out to hang out with some friends, but actually hook up. I'm

23 with my own apartment now, and she comes over almost every day to make out/fuck

40. theyllknow:

My boyfriend and I met at the brothel were I used to work. As a whore.

41. throw7638:

I do not have a lot of confidence, and can never ask girls out. I met my current wife by installing a keystroke logger on her computer, and intercepting facebook messages and chats with her friends until I confirmed she liked me. That way I knew exactly how to approach her. I orchestrated our entire early courtship to my advantage. If she knew she would likely divorce me because I delved deep into her personal life and found out some crazy things about her past.

42. secretthrowawaybla:

I am an active opiate addict. I use every single day. Everyone in my life – even the people closest to me – think that I have been clean for over a year. I'm a good actor and liar, it comes with the territory of addiction. I don't want this, I hate myself, I want to stop more than anything. It's so damn hard.

43. Harlotseeker:

26 year old male, and have "visited" with 30+ escorts over a 4 year period.

44. [deleted]:

I was jumped by a group of gang members a number of years ago. I was hospitalized, wound up with a concussion, broken jaw, 46 stitches and tens of thousands of medical bills I am still unable to pay. I know who all the gang members are and directly recognized one of the assailants and filed a police report. He had an "alibi" and nothing ever came of my case.

I had run in with them again a few years after that and ended up with stitches and no charges sticking to my attackers.

I see these gang members around town still. I get chased out of bars, and there are certain places I don't frequent because I know they may be there. I bought a hand gun just a few short years ago for my own protection and knowing these guys are not just going to let me slide if they run into me again. I carry it if I know I'll be in "problem areas" and neighborhoods where these guys may be.

One night, not too long ago, my girlfriend an I were walking downtown when I noticed a large group of them hanging outside a bar. I told my girlfriend to wait for me at another bar not too far away while I pulled my hat down over my face and put my hood up. I walked across the street to a construction zone where I could keep out of sight and still keep an eye on them. A half hour later two of them came walking across the street passed the construction zone. I popped out drew my gun and fired at them twice, unknowingly missing the first one, but hitting the other in the gut. He keeled over and let out a long groan before falling to the ground. I looked for the fist one and he was laying in the street a few yards away (ducking for cover).

Thinking I had hit them both I ran around the corner pocketed my gun then ran to hide by an over pass a number of blocks away. I texted my girlfriend, she came and met up with me, and we took a cab home which drove by the scene.

The man that I shot is now in a wheel chair, paralyzed from the chest down. They (the police, the gang members, the community) didn't know who shot them, they think it was rival gang members. I still see them around town. They are not any more weary, but I am armed and ready.

I've only told my best friend this story. He told me not to tell anyone else, not only because I could get in trouble, but because it would change peoples perception of me. My girlfriend never really asked what happened that night but she expects me to tell her at some point.

45. blackhawk767:

My mother has multiple sclerosis and her health has deteriorated fast since I have been born. She was gone from being able to walk, to needing a cane, to needing a walker, to complete wheelchair usage, and now completely bedridden. She has a urinary tract infection that is untreatable and is constantly in physical and emotional pain. She takes prescribed medication for depression and bipolar disorder, as well as sleeping pills. Throughout my childhood she has tried to kill herself three times because she wants the pain to stop.

In the middle of the night, I bought something from a dealer and snuck into my house to give it to my mother.

She passed away within 2 hours.

My dad, sisters and brother have no clue.

46. [deleted]:

While on deployment, I killed a man in a coup de grace. The feelings of taking a man's life always weigh a heavy burden on me every day. No one like's hurting people. He had been hit by some of our mobile artillery. While part of me wanted the bastard to be in pain, it wasn't right. My medic was busy with my wounded, and as the officer on duty I took out my .45 and put one in his head. I knew my boys wouldn't say anything. Most just watched, accepted it as a fact of war, and kept walking .I remember throwing up afterwards. I came home and everyone acted like I was a hero. I never felt like more of a sham my entire life.

47. NowYouKnow2006:

I have herpes.

I know that doesn't sound like anything particularly horrible after these devastating tales of incest, rape and other sad/terrible/morally ambiguous situations, but I feel like it has ruined my life. I feel trapped, like I will never find someone who could actually like me enough to see past it.

No one knows. No one would even suspect. I'm quiet and nerdy, keep to myself, keep my nose clean, etc. But I am naive when it comes to guys. Or I was. A boyfriend in college didn't tell me and gave it to me... and then cut off contact when I "realized". I had only lost my virginity a year before that.

I know it seems like nothing in comparison. I knows some would even find it funny. But you have no idea what it's done to me. It's destroyed me. I've considered suicide, its only been this past year where I don't want to walk to a nearby bridge and jump. I feel just... wasted.

Even if you're shy, you still at least have a chance with your crush or someone you like. With this, all my chances have dropped to zero. If you like someone, think about it... would you still like them, want to date them, if they had herpes?

48. secret_sauces:

I have terrible credit. I have debts from 10 years ago that I never paid off. My wife doesn't know.

49. schoolsbelly:

My brother committed suicide in 1994, shortly thereafter I intercepted a letter to my parents from his girlfriend. She was pregnant and wanted them to know and asked if they wanted to be in the babies life. I burned the letter and have never told them. She never contacted them again and I did so many drugs that I buried that secret deep in my subconscious.

50. Data_Points:

I'm white and my wife is half black. I fantasize that she's my slave when we have sex. She thinks I'm the least racist person she's ever known.

12

50 People On 'My Most Embarrassing Sex Story'

1. csoimmpplleyx:

   I was doing missionary with my ex while in high school. We were in the gym and I was so turned on I pulled out and blasted in my own eye. I turned around because my girlfriend had this terrified look of embarrassment on her face only to stare into the angry face of her gym teacher while the spooge dripped down my eye onto my lip. Yeah that was a great day....

2. voice_of_experience:

   In college, I had the best kind of roommate situation: I shared a bathroom with one other guy, and we actually had separate rooms on either side of the bathroom. With that kind of privacy, I could have a lot of loud sex. And I did.

   One night I was sitting on the edge of my bed, with my girlfriend straddling me. It was that great kind of loud, dirty talking sex, and I started to spank her as she rode me. She loved it and kept asking for more, so I spanked harder as we got more and more into it. We got a lot of energy going, and I was smacking her ass pretty damn hard... Until I spanked out of sync with her movement. My hand swung down as her ass moved up, and I ended up missing her entirely and smacking myself in the balls with full force.

3. DavidisGoliath:

So it was my 18th birthday. It was about 12:30am at my house, and my girlfriend of the time and I were laying on the couch watching a movie. My family had gone to bed earlier, and my girlfriend turns her head and says to me "I'm going to give you your birthday present now". We start going at it on the couch, and everything's going well. We're in the spooning position, and there is a blanket covering us up from the waist down. Not too much motion at the time just some good grinding, but I was balls deep in her.

The room suddenly got lighter, but a very natural non-electric light. My eyes look up to see my Mother, Father, and Sister with a birthday cake walking into the room. Singing happy birthday. While I am balls deep in my girlfriend.

TL;DR- My parents sang to me while I was having sex.

4. suckiestbunchofsucks:

Sadly, i was once the foolish friend who walked in. I

came into my buddies room after a night of drinking and him and his girlfriend are laying in bed watching a movie and it went like this. I walk in "yo jay, wanna smoke? ohhh shit! You guys are watching V for Vendetta!?!? i fuckin love this movie" i proceed to grab a seat and talk about how much i love that movie (I'm pretty wasted) and my buddy kinda made hints to get the fuck out that i just didnt pick up on. I eventually just left and he told me the next day... I felt like a fools fool

5. [deleted]:

The first time I attempted deepthroating, boyfriend wanted me to shove his cock down my throat as he came. So like any good girlfriend I did just that when he gave the signal.

I still have no fucking clue what happened, but it turns out semen really burns when it's gushing out your nose.

6. roughingit:

Long time ago. I was having a relationship with someone I shouldn't, so we were sneaking around. We were staying at a friend's place in Colorado, very cool, rustic, fireplaces, make-sure-the doors-are-locked because-there-are-bears kind of place. We were sleeping in separate bedrooms, but in the middle of the night I decide, very unusually for me, to be naughty.

So I psyche myself into thinking I'm this uber sex kitten, put on nothing but a bathrobe and sneak into his room. I slip out of my robe and under the sheets, where he's asleep, and climb on top of him, intending to wake him up mid erection and embody this sexually-adventurous fantasy woman I've created in my head.

Whereupon he wakes up with a start, screams and pushes me off him – and off the bed – because (as he later explains) HE THOUGHT I WAS A BEAR.

This stellar moment was followed by wild confusion on his part (i.e. "what are you doing!?? Why are you here!??" – not exactly my dream scenario) and then me hiding under the bed naked for about 15 minutes in case our friend came looking to see what the yelling and loud thump was about. Although the sound of my ego deflating may have been louder.

TL;DR: Thought I was a sex kitten, he thought I was a bear.

7. antag4123:

Its my college graduation party and me my WHOLE family and a bunch of my friends are celebrating it in my backyard. We are havin a good time drinkin a few beers throughout the day and the later it gets, the more crazy it gets. My friends, some family, and I are all half in the bag. I'm near blackout at this point and all i can really remember is being behind my pool with this girl i was friends with, eating her out.. now in my drunken state i coulda sworn that it was dark enough behind there but when i awoke the next day, my father assured me it wasn't. Not only could he assure that but about 60% of my family could…

TL;DR: I ate out a girl in front of my family.

8. Chile_dawg:

My boyfriend and I did anal..
i pooped on him.
We're married now! ?

9. DaGreatPenguini:

About 10 years ago, I took my GF to a small bed-and-breakfast in Harper's Ferry, VA for the weekend. It was really pretty up there, and being so far away from the city, you could see the Milky Way at night. So we're taking a drive around dusk when my GF spots an old civil-war cemetery; we stop, get out a blanket, a bottle of wine, and my big MagLite flashlight, so we can find our way back to the car.

Needless to say, darkness and drunkenness combine well, and we start boinkin' away on ol' Caleb's burial plot. After I give her the best 20 seconds of her life, we're just laying there naked, enjoying the stars and the warm Summer night, when a car-load full of teenagers pulls up and starts walking through the cemetery (probably to do what we were doing), straight at us. The GF starts to panic because all our clothes are hanging on some headstone four graves away, and we're completely starkers. I tell her to sit back and watch the show.

It's pitch black, so I figure that the kids' eyes haven't yet adjusted to the night. They were ten feet away, obliviously coming straight at us. Just as they were five feet away, I jump up totally naked, turn on my five D-Cell MagLite right into their eyes and yell, "DEA, STOP RIGHT THERE, YOU'RE UNDER ARREST." Well, they scream like a bunch of four-year-olds and bust ass towards their car, flying off in a cloud of dust. That's when I turned to my GF and treated her another 25 seconds of pure ecstasy.

10. Number127:

Late one night when I was in college, I decided to go for a run. I jogged a few miles to a park, and I was feeling pretty good, so I decided to finish by sprinting up a really big steep hill.

I get to the top, now feeling like I'm about to pass out, and as I stagger around panting and gasping, I look up and I see the fleshy shadow of two people tangled up on a blanket a couple feet away, clothes off to the side, looking at me and blinking. I'm so oxygen-deprived that I'm seeing spots, so I can't help but lurch around them like a pervert, taking these enormous rasping breaths, while we stare awkwardly at each other. After ten or fifteen endless seconds I turn around and try to scurry away.

11. Hokuboku:

The one that instantly comes to mind is when my boyfriend and I had sex for the first time after I started using the Nuva Ring. He pulled out after awhile only to find my Nuva Ring caught under his foreskin. For some reason, it reminded me of those hoops that hang out of a bull's nose and I lost it. He's standing there with my birth control danging from his peen and I'm just laughing my ass off. The mental image still makes me chuckle.

12. tippietoe:

My boyfriend stuck his fingers in me and felt the Nuva Ring when we first started dating and thought it was an old condom I forgot about.

13. Fantum49:

When I was in the Army I got very drunk and took a really cute girl back to the barracks. Turns out she was taking me to the female barracks which are (get this) laid out opposite of the males. So after some pretty hot sex, I get up and go to the bathroom, which turns out to be the hallway. I hear the door lock behind me, and the chick is already in the shower, can't hear me knocking. Had to walk back to my room nekkid past a platoon or so of female medics. They laughed, they cheered. I never ever lived that shit down.

14. story1824:

The girlfriend and I had had nothing but bad sex so far in the relationship. Too many hard days at work, too many minor scraps led to some pretty unspectacular sex. It was a doomed relationship, but I wanted to make it work.

So I had her over, made a great dinner and we went

to bed. She fell asleep watching a movie she wanted to watch. So I decided to wake here up with my erotic powers.

I kissed and nibbled and removed her clothes when she finally stirred. She raised her arms above her head and I licked from head to toe and she purred like a kitten. She had a great figure and was now quite 'ready'. As was I. Hard as a rock. I was going to give her the best sex of her life. Songs would be written about this night. My past performances would be forgotten and we would write poems of this.

Anyhow...

I had my arms under her legs and was pointing (literally) toward her when she looked at me with her beautiful blue eyes, glistening skin and fantastic body and said "Oh God, fuck me now".

That was hot.

Super hot. Super duper exciting hot. Uhhh...Too hot.

I came.

Like a fire hydrant, all over her. Covering her chest, neck, chin, forehead and some of the wall. It was colossal.

However, she was furious and stomped to the shower and left. End of relationship.

15. fractalfarmer:

Going down on a guy, and he's getting a bit rough with me. I normally have zero gag reflex and a high pain tolerance, so no problem so far. What I didn't factor in was the food poisoning I thought I had gotten over the previous day. I vomited into my mouth, tried to swallow it back. Unfortunately the taste made me vomit a second time, just as he pulls back and it goes all over my chest, his dick and the sheets. Obviously I was mortified, but he wanted to carry right on fucking me. Turns out it turned him on, big style… I still can't decide whether I am more embarrassed about throwing up or freaked out that it got him going.

16. BrandyAlexander9:

I was orally stimulating my boyfriend at the time when I removed my mouth from the situation for a brief moment and he decided to shoot his man juice up my nose. Between the inferno in my nasal passages and the feeling that I was choking, I was pretty sure I was dying. He literally fell off his bed from laughing so hard.

17. MakeshiftAmante:

I had a guy stop mid-fingering and ask me if I was storing things in my vagina. When I told him, perplexed, that I didn't, he got this look of "AHA! I've caught you in a lie!" and proceeded to exclaim how he knows there is something up there and that I've just confirmed his suspicions that all women store items in their vaginas (I believe he compared us to Kangaroos…) He appeared quite embarrassed for his lengthy diatribe after I explained to him that what he probably felt was my Nuva Ring.

18. WeirdAlLoser:

    He shot all the way to his OWN face/mouth…and I almost fell off the bed from laughter.

19. cylonnumbersix:

    In the middle of taking my boyfriend's virginity, his nose started bleeding right onto my face. I asked him if he was ok, and he said, "yeah, it's just…my nose always bleeds when I get nervous." Cutest. Thing. Ever.

20. drmctesticles:

    Me and the girl I was seeing were real drunk, got back to her mom's house from a booze cruise real late and wanted to have sex. Since her mom wasn't the biggest fan of me (we met when she barged into her daughter's room at 3AM to complain about loud sex noises) we decided to it outside on her front lawn. We did our thing and then immediately passed out, buck naked on her front lawn.

    We ended up being woken up by her neighbor mowing his lawn a few hours later.

21. ForEverythingElse:

    So, I'm a guy. A female friend of mine came to visit me in the middle of the night with a girl she had recently met (and for added spice worked at a sex-hotline).

    They wanted to crash my place for a couple of hours to wait for their ride out of town. They had woken me up so I just sat on my bed chatting with them. My friend came to sit on the bed with me and, before long, there was touching. The other girl was using my computer at the time. Things got pretty heated and after a while the friend saw what was happening and came to join the fun. That's when it turned bad. The thought that this fantasy of every guy might be coming true hit me like a ton of bricks and ... I don't know. I panicked.

I sat up and said: "I'm making noodles. Who wants noodles?" I jumped off the bed and walked straight to the kitchen, feeling their "What the SHIT!?" looks on my back. Then I just stood in the kitchen looking at nothing and ended up making noodles with added tears. They left soon after.

I still sob a little when I think about this.

22. superdillin:

This happened when I was in high school, blow jobs were about as far as I had gotten sexually, and I was REALLY awkward about anything sexual. Also, it was in a time before I had a car or a place of my own, so my boyfriend and I would go to the woods to hang out and fool around. Well, we were standing in the woods at the top of a small hill and I tried to kneel. I stumbled, and started somersaulting backwards down the hill. And he started chasing me with his pants around his ankles. By the time I got to the bottom, I was covered in dirt, twigs, leaves. This did not help my sexual awkwardness. Not one bit.

23. [deleted]:

My g/f and I are swinging and we're at a local club we've never gone to. We're the youngest people there by at least 20 years, we're new, and we're in good shape, so we're the center of attention. My g/f isn't keen on sleeping with any of them, so we're having sex together and attract a few people to watch, which turns into quite a few people in a few minutes. I'm really turned on by the audience, so I'm kind of lost in my own little world when one old guy sidles up behind me and rams his finger in my ass – no lube, no warning, no request. I open my mouth to yell at the fucker and he covers it with his free hand and shushes into my ear like I'm 2. In about 30 seconds, we're in a naked fist fight in the middle of a swingers club.

24. czander:

Was having normal regular sex with my girlfriend of the time. Though I suppose she wasn't exactly wet, neither was I. So were going at it and all of a sudden I feel a kind of pinch. Thought that was weird. Then liquid, so I asked her, do you feel something odd? She thought I came. I hadn't yet, so far it had been really uncomfortable. So I pull out. All of a sudden blood is squirting from just under my penis, all over her vag, stomach, bed etc. I jump up and the blood goes on her floor, chair, myself. I freak the fuck out half yelling "WHAT THE FUCK JUST HAPPENED?!" She freaks out, eyes wide just in shock. I run to the bathroom and just stand in the bath clutching tissues to my penis waiting for the blood to stop. Eventually I get flaccid and the blood flow stops. We immediately google what the shit was going on. I'm clutching my dick through my shorts the entire time as we look for an answer to how this spontaneous satanic blood orgy started and if anyone else has had the same. Turns out you can tear the webbing under your penis. the 'frenulum' i believe. Who knew. Couldn't have sex for 2 weeks and then the next month after that I was too paranoid to do much anyway. Hilarious to look back on though.

TL;DR Broke my dick.

25. easily:

So my friends and I had were throwing a party and I'm upstairs in my room having some vigorous missionary sex with my girlfriend. I guess I was a little under the weather at the time and my nose was running. I wipe my hand on my forearm but my forearm was too sweaty to absorb the snot. I figure, fuck it, it's dark, I'm going to try to finish up without her noticing.

Next thing I know she flinches pretty hard and says "are you okay?" I obviously tell her everything is fine. 10 seconds later she tells me "I...I think you're bleeding!" At first I thought she said she meant it was that time of the month for her and I was pretty nervous that I'd have some serious laundry to do the next day. I jumped up to look and she quickly got up and turned on the lights to look in the mirror. We're still completely naked.

Apparently I had been bleeding on her face for a couple minutes because she looked like a goddamn vampire! Her face and upper chest was covered in my blood.

She runs out of the room naked, kicks three girls out of the adjacent bathroom who are terrified. They look right at me in my room while I'm still naked. I must of just had a confused look on my face then they ran away screaming.

26. iamacynic:

I made my girlfriend, at that time, dinner and it eventually lead to the bedroom. I started to play with her at first and she seemed to enjoy it, but after a couple seconds her face turned from pleasure to discomfort and then near horror. I had been cutting habaneros for dinner and totally forgot to wash my hands with soap. Needless to say i did NOT have sex that night.

27. DocTaco:

Girl pooped on me. We got done having sex (her on top). I stagger to the bathroom to clean up the condom and stuff and there is shit all over the place down there. I panic. Like totally freeze. Slowly I come around to a plan of action:
    Step 1: Confirm that the shit is not mine.
    Step 2: Clean up shit.
    I return to the bedroom and, because I don't know what else to do, I act like everything is fine. She is acting weird, asks, "Is everything OK?" I just play it poker-face style. She goes to the bathroom, comes out, and nothing about it is ever spoken of again.

28. Ihearthuckabees:

When I was a drinker, I finished a bottle of wine off before I began having sex with my boyfriend. I had a crazy screaming and moaning session. (I'm sure my neighbors in that apartment thought I was taping a porno) Afterwards, I got up, put my clothes on, and told him not to tell anyone because I had a boyfriend. I've never been unfaithful and when he told me what I said the next day, I couldn't believe it. He was pretty pissed because I cheated on him with himself.

29. kuzc00:

I was dating this girl with really nice breasts and I really wanted to try titty-fucking her. I got her to agree and quickly found myself plowing those succulent melons. I was approaching climax when she suddenly yelled out that she wanted me to shoot my load in her mouth.

Let's face the facts: I'm no Annie Oakley, Billy the Kid or Wyatt Earp. I'm not a good shot. Honestly, as a man, I'm happy when the lube doesn't end up on my stomach, rather than in the wad of Kleenex.

But, there she was, mouth open and eager. I aimed as best as I could and promptly shot a big load of love juice UP her nose. Now, at this point most women would freak out and shove me off themselves and run like hell to the bathroom.

Instead, I was horrified to see her pinch off her unplugged nostril and SNNNNNGGGGGTTTT (snorted) down my gob of cum. In one quick motion she both frightened and repulsed me.

30. HDMBye:

The first and only time my dad caught me masturbating, I was looking at 5 second clips of women lactating. One was just loading.

I jerked off at the computer in my mom's home office and came on a couple pieces of computer paper. My dad called me from across the house before I could clean up and messily stuffed myself inside my shorts. My younger brother comes in while dad is asking me something and asks what is next to their computer. I fake angrily think, "Oh, oops, I spilled milk right before dad called me." Sprint back to destroy evidence.

Stained the carpet in every bedroom I had growing up with cum. And the areas around each computer in the house had discolored or slightly hardened carpet.

Directed my brother to a porn site when he was in 4th grade. Convinced him it would be cool to add as a link to his website he was making for class. Much trouble. For me.

Shit in a ziploc bag and hid it in the bathroom when I was 10. Checked on it every once in a while to see how it changed. Eventually was worried of being found out and threw it away.

Peed in a water bottle the year before with similar results. Oh wait, my mom found it in my desk.

Parents' friend stayed with us for the week when I was 11. One night she didn't come home and I was up working late on a project. Shaved off my pubes with her razor and wore her panties. I think she knew somehow?

At 9, I covered myself head to toe with toilet paper while I had a boner and got in the shower. It was Christmas Eve. My grandma kept thinking I had dandruff or something at church, picking bits of white out of my hair.

31. PoonRaider:

Doggie style. She clearly hadn't wiped well enough earlier that day. I immediately turned off the light next to my bed so I wouldn't have to look at it but that didn't kill the smell. Fucking really hard and fast helped take my mind off it. Still a win in my mind. She was 5'6" and 110lbs and a former model. Even hot chicks can be really gross.

32. [deleted]:

Oh and then there's st patty's day, 3rd year of college, got insanely drunk, picked up a girl at a bar, had sex with her on her nasty shared dorm bathroom floor and proceeded to puke into her vagina as I was going down on her.

33. thepooper:

So cute girl in my philosophy class, apparently tells one of my friends that she's crushing on me and needs to get laid, so we all go out together to a bar one night and I end up taking her home. Girl is a freak, straight up. First time we fuck, I get the condom on and all that, but after that she just wants to raw dog it – mistake #1. She gets on top and after riding for a little starts trying to stick it in her ass, which is so tight that the weight just bends my dick all outta shape for a while until finally we do a little >((which turns out to be quite pleasant. Anyway, pass out. Wake up. Decide to give it another go, doggy style this time. After a few minutes of vag soufflee, I go for anal again (mistake #2), and this time I slide right on in. I start going at it, she's moaning and all that, and then all of a sudden it starts to smell bad. Really bad. I look down and the inside of her butt cheeks are gooey and brown. Gag. I pull out, and she immediately starts farting. "Oops", she says, and excuses herself to the bathroom. I look down, and my poor shlong seems a little darker than usual. I grab a tissue, do a quick wipe, and turn it around. Feces. I almost vomited.

34. story2466:

Thai food apparently got my girlfriend in the mood. Buying her dinner at Sawatdee worked great, and we were experimenting with deep-throating later that night. She was more eager than talented, though and threw up. Not wanting to actually puke all over me, I think she tried to hold it in by pushing harder. That mostly worked, but it killed the mood.

I woke up in the middle of the night with my first and only urinary tract infection. Holy shit is that painful. She brought me to the hospital and watched while the ER doctor stuck an oversized q-tip up my urethra. I married that girl.

The doctor told me (though not in exactly these

words): If someone pukes on your dick, pee right away. It flushes out the bad stuff. This advice has served me well as my eager wife continues to learn.

35. [deleted]:

None, because I'm still a 22 year old virgin. Now I made myself sad.

36. recl00se:

Mine was more funny than embarrassing. I have muscular dystrophy, in a wheelchair, and don't have the best upper body strength/balance. However, my gf at the time and I figure out that it's a lot of fun to fuck sitting up... with me sitting somewhat indian style.

We get more and more into it... bouncing harder and harder... when all of a sudden the bouncing unlocks one of my elbows which im using to keep myself upright.

My upper body collapses to one side. I let out a sound that my girlfriend said sounded exactly like R2D2 screaming... as i go flying off the side of the bed.

The only thing keeping the lower half of my body on the bed was my girlfriends vagina. Which after a lot of laughing and discussion, she removed herself from me and the remainder of my body fell of the bed.

37. baldric01:

I went on a date with a girl after we met at a speed dating event. I ended up back up at her place, not sure how we got there as I was pretty drunk at that stage. We shagged and a few hours later I needed a piss so got out of bed and started to piss across the room before I realised I hadn't made it to the toilet (first and last time I've done that… i think)

I staggered down the hallway naked to the toilet. I went back to her room and started spooning thinking it was time for another round. I was slightly surprised to hear a guy say "uh I think you might have the wrong room buddy". I had gone into the wrong room and ended up in bed with a guy and his girlfriend. The crazy thing was that he seemed so relaxed about it! Let's just say I didn't hang around for breakfast the following day.

38. HDMBye:

we are going at it and the condom breaks. I think she just gave me head afterwards. She leaves on a trip for a month and I go to pick her up from the airport. When we hug, it is electric. We are super eager to have sex but the airport is 2 hours from home (cheaper flight in). We are touching and basically driving dangerously the whole way. Road head for the last 45 minutes of it. My parents are out of town and their house is 15 minutes closer than mine. I pull in, we run out back to the pool and strip down in their secluded back yard.

We begin going at it on the tile next to the pool. I notice this itch hitting the head of my penis. Wtf? I keeps happening when I am hitting inside of her a certain way.I stop. She asks, "What is it?" I just say, "Hold on," and go in with my fingers (I have small hands).

She can tell I am searching and not trying to pleasure her and I just say again, "Wait a second." Groping with my fingers, I feel it. The thing, the itchy thing. I get two fingers around it in a scissor grip and pull it out. "It" is the reservoir tip from the last time we had sex, a month ago, covered in discharge over the past month, etc. This didn't stop us o kill the mood but it was nasty. I just considered it a primate grooming thing.

39. [deleted]:

I'm a pretty pale dude so I decided to do some tanning a few days before my first time. Needless to say I overdid it. I was burnt like you wouldn't believe. When we got around to having sex, all the sweat and body contact caused pretty much my entire outer layer of skin to peel off during the sex. Jesus it was horrible…

40. [deleted]:

I was ABSOLUTELY pissed and high. I meet this girl, we get friendly, we drink so much now that I barely notice consciousness. She pulls my pants down —> fellatio. She then stands up and takes her own pants off. When she sits on me and I casually look over her shoulder, I realise that we had never left the pub we were at and was staring into the eyes of a very drunken and —to my horror— masturbating Irishman. Pub owner comes along = banned.

41. dante428:

I tried sex in the shower one time. I picked the girl up and had her pressed against the shower wall as I was pounding away. I soon lost traction with my feet, dropped her on her ass, and fell backwards out of the shower tearing the shower curtain down in the process. She told me I had a look of shock after it happened. Luckily only my pride was hurt.

42. 40ozToFreedom:

One time I was having sex with this girl and I was on top so I had my head to her left and breathed really hard and ended up snorting a booger in her hair. I kind of freaked out but I ended up "stroking" her hair to get it out. It was funny. I don't think she ever knew.

43. Curiosity27:

Worst I can think of happened with my ex.

He was a douche, and pretty awful at sex because he only cared about his needs and not mine.

After pretending to be completely disgusted by the thought of anal sex for about a year he did a 180 and decided he wanted to try it. After that, that was pretty much the only type of sex he wanted.

One day we start making out and stuff and it leads to sex. But, something had not been agreeing with my stomach that day and I hadn't had a chance to "prepare" either. I told him no, it was not a good idea right now and that he would get a lot more than he bargained for. He kept insisting and I finally flat out told him that I'd had diarrhea all day and still felt sick.

Like the douche he was, though, he either didn't care, or didn't understand what would happen. Let's just say, he put it in, and as much as I tried to stop anything bad from happening, I couldn't help it and pooped all over him. I was so embarrassed that I was sobbing, and then I was pissed because he didn't listen. He even had the gall to get mad at me as though I'd done it on purpose. Needless to say, we never had that type of sex again.

Actually now looking back on this it makes me laugh, because he most definitely got what he deserved.

44. [deleted]:

This guy and I were newly dating and one night having sex on the floor. So we are getting it on and all of the sudden blood starts gushing out everywhere. I thought that it was my period until he pulled it out and I quickly realized that his foreskin had kind of.... ripped. (He was uncut) Turns out he had contracted some kind of yeast infection (Apparently this can happen) which caused the skin to become tender and TEAR. The carpet was white and became SATURATED in blood. The worst part was we were 16 and his mom was outside honking the horn while his penis was bleeding everywhere.

45. [deleted]:

Nothing too bad, but my SO's girlfriend's mom walked in on us one time having sex.

This is the words spoken.
"OMG I'm Sorry!"
"Mom, GET OUT!!"
"Sorry!"
door slams
"I just wanted to let you know that I made quiche."
"The quiche can wait damnit!"

46. ImNotBecky:

Me and my then boyfriend were goin at it and out of nowhere...
  Him: E.T. phone home.
  Me: uuuhhhh what??
  Him: What? Why did you stop?
  Me: You just said "E.T. phone home" as you poked my boob.
  Him: No I didn't... i said it in my head but not out loud.
  Me: Yeah. You did. I heard it.
  Him: oh...

47. DadReadsReddit:

His mother and her boyfriend walked in on us so many times it wasn't even embarrassing anymore... Until she almost saw me naked, tied to his computer chair with his gamecube controllers.

48. tempexperiment:

I stuck my finger up my ex's butt and upon pulling it out I discovered it was covered with little white worms.

49. SlimDuncan:

Girl's leaving my room.
    She sees my magic cards.

50. inkathebadger:

My SO and I were going at it with me on top leaving his hands to rub, tickle, scratch and whatever.

Well in the heat of the moment he reaches up and give my nipples a squeeze. Usually something I like but at that moment was not expecting it so I squeaked and jerked back, he rocked forward with me and half a second later I jerked forward again and we cracked skulls. Hard. I mean we both had tears in our eyes and I was dizzy and couldn't stand up for a few minutes. I was almost worried we gave ourselves concussions.

After we recovered enough to be able to see straight, what does he think? This will make a funny Facebook post! My mom saw it and commented on it.

13

21 People Confess The Big Secrets That Could Destroy Their Relationship

Every one of us is hiding something, a secret we're too afraid to tell others, because we feel like it's too big or they couldn't accept us for it. These secrets become a part of us, little boxes locked away that we protect. On reddit, 21 people confessed the secrets they're hiding — either from their significant other or from the world — and they range from contemptuous to heartbreaking.

---

1. BX8

I was the reason my girlfriend's brother killed himself.
 One night we were sleeping and around 3 A.M., her brother called asking to speak with my girlfriend (much like he did all the time — but usually not this late).

Quick back story, the brother was heavily addicted to drugs and alcohol.

Anyways, so he called in a drunken mess asking to speak with her sister (my girlfriend) and I'm usually very calm but for some reason, maybe I woke up on the wrong side of the bed or whatever it may be. I shouted, "Dude, she's fucking sleeping, she doesn't want to speak with you," and hung up.

Well, turns out he broke up with the "love of his life." Later that night, he threw himself off an overpass onto the freeway and died.

What sucks is, several weeks/months later she said to me, "I wish he had called me. Maybe he would still be alive today."

So in the end, I never told my GF or anyone that he had called.

2. UHLIZUBETH

I really, really dislike his mother. I've never really disliked somebody until her. Sometimes I think about leaving him because I fear having her as a mother in law someday.

3. SALLY-SOMF

I am heavily addicted to snorting heroin. When we first met she found out, but at the time she was just a friend of a friend, i didnt care how she felt about it. After a month or so we bumped into each other and I asked her out.

One of the first things she asked me on our first date was if I was sober or not. I lied and said I was "clean."

We've been together for about 9 months now and I've been using for 3 years. The main reason she wants me clean is because she doesn't want to find me dead one day. She even kicked a pill popping friend out of her house when she offered me some. I love her and I don't want to lose her and I should get clean for my own good but I just can't stop.

4. SOUPSPA

I masturbate furiously to other women when I'm mad at her. I don't think she'd dump me for that but she'd get really mad.

5. TISKEL

I just started dating a sweet, intelligent, hilarious, super-dorky guy. He is Christian and (I think) quite conservative. His family is quite traditional too. He held on to his virginity for a while and is definitely not the kind of guy who sleeps around.

He is perfect for me in every way. I don't want to fuck this up.

I have no idea how to tell him I worked as a prostitute for a while, and it's not something I can keep from him with a clear conscience if this goes anywhere. Pretty sure it will completely change his view on me.

6. SCROTUM_NACHOS

I don't love her but rent.
It makes me sad at myself.

7. MANGY_MOOSE

I had sex with a prostitute.
It's a LDR and I don't love her anymore. Stationed in Korea, and tired of her sneaking around with her ex, and finding out she has feelings for other guys. I'm breaking up with her ASAP. I don't regret it.

8. VENTMONST3R

I only proposed to her because if I didn't she would have killed herself.

9. PRETTYFORARIOT

When I was 14, I (25f) was raped and at some point orgasmed for the first time in my life. Since then, I've never been able to cum with another person in the room, but can make myself cum with no problems over and over again. I used to not be able to be touched by anyone without freaking out, but I got help from a therapist and am left with this final wall.
My boyfriend doesn't know that every orgasm he thinks I've had is fake. I really want to tell him because it's the only thing I've ever lied to him about and I think it's a big nasty lie, however every guy I've ever told has either left me because I'm broken or rubs my clit raw trying to be my white knight and save me from my problem without really trying to find out what I

really need to overcome it. Doesn't matter how much I love the sex, apparently I'm only worthwhile if a guy can make me cum.

I'm pretty sure after the initial shock and anger at me lying for this past year, he'll still love me, but in the back of my head a tiny little voice tells me that when I tell him he's going to leave me because I'm broken or that my inability to cum is going to ruin our pretty fantastic sex life. Or that I'll hurt him. I think that's my biggest worry.

10. JONATHAN88876

I love mine, and she's an amazing person, but if my ex were to kiss me right now, it would all be over.

11. OHCHICK

That I am still legally married.

I don't have the money for a divorce, left my husband 4 years ago in another state, and wouldn't know how to contact him if I wanted to.

12. DOUBLEDICKING

I was going on a few dates with various people, living the single life, and accidentally ended up exclusively dating two of them. I now have two SOs, both are totally unaware of the other. This has been going on for about two months.

I can't choose between them. Oh man, I'm a bad person.

13. MILDLYENTHUSIASTIC

That I'm not gay. I love my SO and would never cheat, but I'm very much straight and miss having [heterosexual] sex.

14. HIMYNAMEISERICA

I try not to hide anything from him, but I can never come out and say that I don't like his best friend, that's also a woman. I know that she has feelings for him, but he will never realize it. Everyone can see it but him…and that kills me.

15. BLACKSTACKS

I cheated on her with the person she hates the most in the world.. Regretted it immediately and eventually cut off all ties with the person. It's awkward because we still see her around and have to act like nothing happened.

16. FLAMEDFUCKFACE

That I make the most of my money from distributing illegal substances.

17. HUR_HUR_BOOBS

I desperately want to leave for the other side of the earth for a year or two. Not because of her but because for over a quarter of a century I haven't moved from this place and I need to get out before I have no chance of fulfilling this dream of mine (Australia) anymore because of real life obligations.

But of course 6 months before finalizing shit I have to fall in love with most intelligent, caring, patient, sexy blond bombshell I've ever seen.

18. N8JS

I am not in a relationship because i feel that my secret would be relationship ending. I cannot under any circumstances imagine someone accepting me, especially in a long-term relationship due to the nature of my past. My father molested my sister and beat me up all the time. I feel like any one who hears that thinks, "Why would I want the son of a pedophile to be the father of my child?"

Every time I'm paralyzed in fear when I go to meet anyone new...because I feel like I'm lying or hiding. I just cant get away from it, and it haunts me everywhere i go. I feel like a monstrosity most days, so I don't meet new people and I feel like I have to just inevitably accept the life of a hermit.

19. QUESTIONABLE2

Secret credit card. We agreed to never rack up bills but I did. I feel like a cheating hypocrite. Also, currently unemployed so no way to pay it off secretly.

20. NOTME_12345

I am sleeping with her sister on a regular basis. There, I told someone.

21. MAIN_ENIGMA

She's not as good in sex as my ex was. My ex was wild and was into trying a lot of things. If she wants something, she just asks. She also doesn't have a problem letting me know when she wants to fuck.

My current SO has a "I'm not a slut so I'm not gonna do that" mentality. Our sex life is pretty average and have been the same routine for the last 2 and a half years.

30 Relationship Red Flags That Most People Ignore

Have you ever disliked something about your significant other but not wanted to admit it to yourself? I've had that happen before. I've thought, "I'm really not sufficiently physically attracted to this person, if I'm honest with myself," or, "I'm not very proud of my girlfriend and don't really want to take her around my friends." Hopefully you would never have those red flags. Some are more subtle, but just as toxic to a relationship. Looking through this r/AskReddit thread I found some good examples of relationship red flags many of us might overlook.

---

1. DWARF-SHORTAGE

    When they never apologize or takes responsibility for bad behavior

2. CAPTIAN_COCKSMITH

Being dependent on you to be happy or entertained. That is the calling card of a needy, insecure and possibly crazy piece of baggage.

3. HELODRIVER87

I needed this tip back in college. Anytime I went out to do anything, she guilted me into bringing her. Didn't matter who it was or what we were doing, she had to be there. If I ever went anywhere without her, she sulked and got pissed. That relationship went on about 6 months longer than it should have.

4. DEILAN

One of the red flags I totally ignored in a past relationship is that I didn't really like any of her friends. If you don't like the people your SO chooses to hang out with, you probably should reevaluate things.

5. CHRONO1465

If s/he seems displeased any time you go hang out with your friends rather than spending time with him/her, it could be a sign of bigger issues down the road. I've seen many relationships deteriorate quickly, to the point where the significant other eventually unfriended nearly all their boyfriend's contacts on Facebook, saying "You've got me, so you have no reason to need anybody else." Obviously it doesn't get this far on the first date, but it's a very slippery slope, so watch out for warning signs.

6. SWEETDUCKLING

When they don't want you to be friends with their friends.

7. HEYIRV88

When all their exs are 'crazy', nope, common denominator is you dickhead.

8. LITERALLYOUTTOLUNCH

In the beginning stages – when they complain about their ex. It isn't easy to build a new relationship on the ashes of an old one.

9. SKUMFCUK

Holds on to literally everything and brings up stuff you said months ago, even if you forgot saying it. That scorekeeping stuff gets old really fast, especially when you don't remember if it's even accurate or not.

10. GHITIT

If the guy says "you don't really want to date me – I'm an asshole", believe him.

11. BRITTYGREE

Serial monogamy. If someone just got out of a relationship and starts dating you right away, chances are that they don't actually like you, but they like having someone in general. They're just with you to fill a void, and the second you break up, they'll be onto the next person.

12. HEEBS387

When the time you spend with your SO starts being talked about as if there is a minimum requirement per week. Once you feel like you need a time card, its time to punch out.

13. HOMERBM

Any time the relationship needs to be kept secret, there is a problem in there somewhere. I've fallen for it twice and learned my lesson!

14. ANDROMEDE

I'd say it's a bad sign if the person treats a difference in your relationship preferences as a wrong/right situation, rather than as a difference in preference. For example, if you want to see your significant other every single day but they don't feel the same, that doesn't make you "clingy," it means you need to either compromise or consider that you just might not be compatible. It makes me really sad to see people get convinced they are wrong when they just have a preference. And I think those who try to convince others that their preference is the "right" way to behave in a relationship are manipulative jerks.

15. DARTHMELONLORD

If they don't have any hobbies. This is a serious red flag because these people tend to be really clingy and jealous.

16. GOOSIEGIRL

This more applies to those seriously considering marrying their current SO – Having different religions, views on children, or what is important to save for. Those may seem obvious, but they're easy to ignore. Oh, you're Catholic and I'm Baptist? Cool, we're both Christians, right? Or I'm a lapsed Jew and you're atheist, great! Yeah, until one of them decides hey, our future kids must be raised in this religion (or none at all). Hey, I want three kids, he really wants one. That may seem like an "oh we'll figure it out when we get there thing" but that's too late! Who will be the main caregiver? Will that person still work full-time? The spender/saver issue gets a lot of attention, but what about what you actually want to spend your money on? Would you rather have a smaller house and more vacations? Or a fantastic kitchen and always buy used cars? Those types of things seem to be often overlooked.

17. ANGRYASTROCYTE

If they consistently make you their last priority, or simply an afterthought.

18. THEBLOODOFTHEMATADOR

If they want to make "rules" about things you do that they have no business making rules about (i.e., where you go, who you talk to or hang out with), or want to control things like how you dress and how you wear your hair. If they can't stop talking about their ex, they're probably not over it yet and nothing is crappier than a relationship where the ex's ghost is always chilling in the corner. If they constantly have to have their hands on you in public. It's weird and territorial.

19. J517925

When they treat their mother poorly. I dated a guy who seemed really kind at first, and I then met his family. His mother was one of the nicest, sweetest people I've ever met. I noticed little things, like when she would ask him a simple question he would totally snap at her in a really nasty way for no apparent reason. He turned out to be an awful partner, in more ways than one. However, I couldn't get over how he treated his mother and the way he started to treat me after the 'honeymoon phase' was over.

20. ANALOGART

Mood swings. If you're noticing mood swings that come without warning and confuse the shit out of you, run. The earlier this starts, the faster you run.

21. TIMEAISIS

If they make you annoyed/unhappy more often than they make you happy.

22. CORSETING

When EVERY date involves spending money. Alot of women have this need for money to be spent for it to "count as a date".... Even if you're going half and half. A real couple needs to be able to enjoy those do-nothing, watch a movie kind of nights... That's what real day to day married /future life will be like!

23. EXCUSEMESIR_

If all of your friends, or your trusted family members, hate your boyfriend/girlfriend. Often, they can see things about your SO that you can't.

24. CKBROWN10

Longer than usual breaks from sex or other physical contact. I was naive enough to think it was actually work related.

25. WAYNEBRADYSWORLD

When the relationship starts while the SO is already in a relationship. Seriously, it will happen to you next.

26. ALIAS28

I always look at the Purse. This may sound absurd, but

I've found that girls with large purses usually carry a lot of baggage. No pun intended.

27. PERFECT1ONOWNS

If you text your SO and never really respond in a reasonable time, but when they are with you, they are CONSTANTLY on their phone. Thats a serious red flag.

28. TOBIOSFUNKE

If they have a history of cheating on their SO. "Once a cheater always a cheater."

29. MRG06

Unnecessary jealousy. A friend of mine is in a long distance relationship with a girl in the Philippines and some girl on his Facebook wished him a happy birthday. The Philippines girl called him not even 5 mins after it happened and bitched him out.

30. JAY2THEMELLOW

Everyone seems to be posting obvious ones… here's some I've found to be not-so-obvious… 1.) If they need excessive attention from the opposite sex. I'm not saying they can't talk to guys/girls or have guy/girl friends, but if they allow them to flirt and spend an excessive amount of time messaging/hanging out with them… that's red flag #1. 2.) If they do things specifically because you're not there; things they wouldn't do if you were, especially if they don't tell you about it... that's red flag #2. This could be a sign they're ok with doing things behind your back. 3.) If they're influenced by their friends and/or have friends who are bad influences... this could be verrrrrry dangerous... red flag #3.

15

40 People Confess The Little Things They Find Incredibly Sexy

As is its wont, Reddit's "Ask Reddit" section polled Reddit users on the everyday qualities they find sexy in a mate. Usually I'm not that into these threads, but I found this one uniquely fascinating — as a look into the parts of sexuality we don't talk about. Sure, people say things we all know like "confidence" (seriously, does anyone not find that sexy?), but others respond with things like "back dimples" and "messy hair buns." One guy said he really likes it when girls clean their glasses. Clearly I'm way behind on my fetishes.

Some of them are a little creepy, but most are just honest and kind of sweet, a look at what romance and attraction actually are. Human sexuality is everything you can think, strange perverse and often beautiful. It's what makes love interesting.

These are 40 of the best responses. For more, check out r/AskReddit:

1. THREEGIGS

   When a girl wears one of my button-down dress shirts and panties.
   Sexiness isn't showing off everything you've got. Instead, it's about hiding just enough so that when imagination takes over, the result is even better than reality.

2. GULALUSC

   I think it's attractive when a cute girl is driving a shitty car.

3. BRI-BIRD

   When you whisper in my ear. Could be the ingredients to a pot pie and I'm still turned on.

4. MCDANGERTAIL

   I love to see people doing delicate, concentration-requiring work with their hands. Puzzles, models, art, suturing, even some gaming – I think it's the combination of steadiness and interest.

5. SKEETRONIC

When they go out of their way to include (or acknowledge) me. Even just to say hi. That doesn't happen often enough.

6. GO-WITH-THE-FLO

Gentlemanly behaviour. The first night I got with my current boyfriend, we were walking home from a club with a group of friends who I lived near but got left behind due to us stopping to mack in the streets for a little bit. I wasn't about to walk 45 minutes home by myself while he went off home in the opposite direction, so he offered up his place but was so insistent that it was up to me and that he wasn't trying to get in my pants. Then he carried my heels for me until we got to his place, gave me some clothes to wear as pyjamas, and assured me that he could sleep on the floor.

In a very short span of time, I went from "I don't want to have a one-night stand with this guy who I barely even know" to "Do dirty things to me all night long, please."

7. THEDAVIDBOWIE

Confidence.

8. RIZFACE

Guys with beards that look like they're 5 o'clock shadow, but they're meant to look like that.

9. LIL-PRAYING-MANTIS

Beards. Sweet Jesus it just turns me on so much. My boyfriend has one and it just frames his jaw and makes him look so manly. Sometimes he gets lazy and just let's it all go (then looks like a murderer) and when he cleans it up into a nice thick chinstrap, I wanna drop my panties and throw them across the room.

10. FARTSARECOLONKISSES

Breathing on my neck….or even getting close to it.

11. SLEEPYCHELSEA

When they're talking with their friends, or whoever it may be, across the room and in the middle of that conversation they look up over at you and smile.

12. ZSAFREIGH

When a girl has a hair-tie on her wrist, and then she grabs her hair, twirling and twisting it up, rolling the hair-tie off her wrist to secure the hair in place. There's something really sexy about that for some reason.

13. SHAZAMTADA

A very well spoken lady…Sets the mood.

14. SANGIVSTHEWORLD

In a physical way: Naked shoulders, that just makes me crazy. In the other way around: A girl able to challenge in an argument without shouting or being a prick.

15. SECRETLY_STALKS_YOU

When a girl puts on her thinking face. As in, she crinkles up her brow and tilts her head and looks to the top right of her vision while putting her nail-polished hand on her chin pensively.

16. DOOMSKANK

Being comfortable around me. Knowing a girl is happy and willing to be around me and be herself is a huge deal. Bonus points if she's emotionally secure.

17. JOSHUA_SEED

Eye contact.

18. JARGONPHAT

A woman's clavicle. Don't know why, but I've always found it sexy.
   To be fair, it DOES generally point one's eyesight to the middle of a woman's chest… And to be more fair (and perhaps equally shallow), being able to SEE a woman's clavicle generally indicates a BMI that does not suggest a sedentary life style.

19. JAY_DEE_

I recently found out my best friends boyfriend has this

thing about watching girls eat. He couldn't get hard before he told her. But now that she knows he brings over snacks and candy and he's now jizzed twice in his pants by watching her eat.

20. STICKLEYMAN

Girls with ponytails and baseball caps. I love that sporty, All-American look. There's something simultaneously innocent and sexy about it. You know, where the ponytail comes out the back of the hat, like this. I love that.

21. BULKYPANDA

Deep voices. I love to hear a man with a deep voice speak softly.

22. MCPOODERTON82

When my gf gives me "that look". You know, that sexy lip biting, I want you right now look.

23. NATHAND3

One is when my girlfriend talks sports. When we starte dating she hated sports. Over the past year or so she's gotten really into baseball and football. I'm a sportswriter so sports are an important part of my life. When she tells me how many homeruns Chris Davis has, or talks about a players OBP and who struggles to hit a curve…oh baby.

24. NINE-FOOT-BANANA

    Being wanted.

25. VIDERHALET

    When a girl places both hands flat on my chest. Which can turn weird if its a security person just stopping me from getting in somewhere. Then again who would use two hands in that case? Two hands is sign language speak for the D!

26. KUMANOKI

    It's when my wife gets excited about something. Not in a sexual way, just when her eyes light up and something really catches her attention and she engages an idea or project or something. I kind of feed off of that energy and it gives me the happy pants.

27. IFIFORGETTHISITSGONE

    Lazy day clothes. The shirt and no bra look is great.

28. ITS_CALLED_AN_AGLET

    Glasses are also a turn on. It makes a girl look more…. sophisticated and intelligent. It's sexy.

29. KULUS

    When I am minding my own business (reading or

watching tv) and I look up and catch my boyfriend watching me.

30. MOUMOUREN

Innocence. Seeing an innocent wallflower clueless of her surroundings is like a breath of fresh air.

31. PANICINBABYLON

Scruffy guys all dressed up. When my floppy haired, manly man is in a suit and tie it takes us multiple attempts to get out the door.

32. JASONMAC42

On a woman- back dimples. Some girls have, some don't. Two symmetrical indentations on the hard, flat lower back. Yowza.

33. QUIRKY_QUINN

A guy with an awesome sense of humour. Drives me crazy.

34. ONWARDAGAIN

Girls with a bike! Or a backpack! Or clothes that say "My fun doesn't come from a TV, it makes me sweat". By which I mean it's more like kayaking or canoeing or hiking or biking or jogging.

35. MRX009

Helplessness. I'm not a rapist I swear.

36. CHARDENAY

That line of hair that some guys have that goes from his belly button down! Don't know if there is a name for it but when it's visible above a good pair of jeans – sexy – oooo la la!!

37. ANTICOCKBLOCKMISSLE

Tits.

38. AUDHUMBLA

When women can have an interesting conversation with me. Not just the "OMG those guys broke up, and he went crazy because she slept with that dude" bullshit that girls of my age often talk about. The few times I can actually have an interesting conversation I often can't get that girl out of my head for days.

39. GRUMPYCATT

When we're walking together and he puts his hand on the small of my back.

40. FRAGILEPENGUIN

When a guy unintentionally shows how strong he is. Pick me up and throw me over your shoulder like a rag doll. My pants will fall right off.

Readers, what are the little things that turn you on?

Which of these qualities also attract you to another person and which do you find creepy and repulsive? Sound off in the comments. Let's talk about sex.

THOUGHT CATALOG Books

Thought Catalog Books is a publishing house owned by The Thought & Expression Company, an independent media group based in Brooklyn, NY. Founded in 2010, we are committed to facilitating thought and expression. We exist to help people become better communicators and listeners in order to engender a more exciting, attentive, and imaginative world.

Visit us on the web at *www.thoughtcatalogbooks.com*.

Thought Catalog Books is powered by Collective World, a community of creatives and writers from all over the globe. Join us at *www.collective.world* to connect with interesting people, find inspiration, and share your talent.

Printed in Great Britain
by Amazon

55666258R00179